Heath Hens *(Tympanuchus cupido cupido)*
by Louis Agassiz Fuertes (1874–1927)

Louis Agassiz Fuertes painted these Heath Hens "booming" at their last known breeding site, the heathlands of Martha's Vineyard, Massachusetts. An eastern subspecies of the Greater Prairie-Chicken, the Heath Hen once ranged along the coastal plain from southern Maine to Maryland and Virginia, but due to habitat loss, overharvesting, and predation, it became extinct in the 1930s.

Grasslands of Northeastern North America

Ecology and Conservation of Native and Agricultural Landscapes

Peter D. Vickery
and Peter W. Dunwiddie
editors

Center for Biological Conservation
Massachusetts Audubon Society

Copyright © 1997 by Massachusetts Audubon Society

Published by the Massachusetts Audubon Society,
208 South Great Road, Lincoln, Massachusetts 01773

All rights reserved. Except for brief excerpts in connection with reviews or scholarly analysis, no part of this publication may be reproduced, stored in a retrieval system, or transmitted, in any form or by any means, electronic, mechanical, photocopying, recording, or otherwise, without prior permission in writing from the Massachusetts Audubon Society.

Library of Congress Cataloging-in-Publication Data

Grasslands of northeastern North America: ecology and conservation of native and agricultural landscapes / Peter D. Vickery and Peter W. Dunwiddie, editors.
    p.   cm.

Papers presented at a conference held at the
University of Massachusetts, in April 1994.
   Includes bibliographical references and index.
ISBN 0-932691-25-0

1. Grassland ecology—Northeastern States—Congresses.
   1. Grassland conservation—Northeastern States—Congresses.
   1. Habitat conservation—Northeastern States—Congresses.

I. Vickery, Peter D., 1949-   II. Dunwiddie, Peter W., 1953-

QH104.5.N58G735   1997   97-19636
577.4'0974—dc21   CIP

Dedication

This book is dedicated to several people who have been important in our lives: to Eugenia Vickery, a mother whose strong love for the soil and its plants, and whose unrelenting pursuit of excellence, helped foster an unshakable commitment to conservation and "to doing good science"; to our wives, Barbara Vickery and Liz Bell, for their strong support, guidance, and love; to our children, Simon and Gabe Vickery and Louisa Dunwiddie; and to all children and future generations. We hope this book will provide greater impetus to conserve and intelligently manage the special habitats that we have had the good fortune to study, burn, and cherish.

Acknowledgments

Numerous people have contributed to the development and production of this book, and we warmly acknowledge their guidance and help. Discussions with many different people, most notably Barbara Vickery, Tom Chase, Tim Simmons, Lisa Vernegaard, Liz Bell, Chris Leahy, George Jacobson, Jr., Mac Hunter, Greg Shriver, Scott Melvin, Susan Antenen, the Martha's Vineyard Sandplain Grassland Working Group, Andrea Stevens, Jeanne Anderson, Julie Henderson, and Nancy Sferra, have stimulated our interest in the origins and management of grassland habitats. James R. Herkert, Greg Shriver, and Steven A. Woods reviewed and improved several of the manuscripts. Critical logistic support was provided by Barbara Vickery, Liz Bell, Diana Coonradt, and Jaime Bodge. Curt Griffin was especially helpful in overseeing the review of several manuscripts and in facilitating and hosting the Grassland Conference on the University of Massachusetts campus in April 1994.

We thank our colleagues at the Massachusetts Audubon Society—Betty Graham, Ann Hecker, Vanessa Rule, Holly Prees, and especially Chris Leahy—who helped produce this book and Barry Van Dusen who did a wonderful job designing the book.

The Upland Sandpiper on the front cover was painted by Lars Jonsson and is reprinted with his permission. It appeared originally in *The Nature of Massachusetts* (C. W. Leahy, J. H. Mitchell, and T. Conuel III; Massachusetts Audubon Society 1996). The Heath Hen was painted by Louis Agassiz Fuertes and originally appeared in *Birds of Massachusetts and Other New England States* (vol. 2, E. H. Forbush; Massachusetts Department of Agriculture 1925). The Fuertes painting is owned by the Commonwealth of Massachusetts and is reprinted here with the permission and assistance of the Massachusetts State Archives. The line drawings of grassland birds that appear in the text were provided by Barry Van Dusen. We extend our warmest thanks to Jonsson, Van Dusen, and the Massachusetts State Archives for their contributions.

Finally, we especially appreciate the remarkable job Elizabeth C. Pierson has done. Her thoughtful editing, meticulous copy-editing, and unfailing good humor have been essential to the completion of this project.

Contents

Introduction 1
Peter D. Vickery and Peter W. Dunwiddie

Thoughts on the Biogeography 15
of Grassland Plants in New England
Leslie J. Mehrhoff

History of Vegetation and Fire on the 25
Pineo Ridge Pine Grassland Barrens of
Washington County, Maine
J. Chris Winne

Paleoecology and Historical Ecology 53
of an Extensive Bluejoint Reedgrass Grassland
in Coastal Eastern Maine
Ann C. Dieffenbacher-Krall

The Role of Nutrient-level Control in 69
Maintaining and Restoring Lowland Heaths:
British and Northern European Techniques of
Potential Application in Northeastern North America
Wesley N. Tiffney, Jr.

Coming Full Circle: 79
Restoring Sandplain Grassland
Communities in the State Forest on
Martha's Vineyard, Massachusetts
William H. Rivers

Vegetation Management in Coastal 85
Grasslands on Nantucket Island, Massachusetts:
Effects of Burning and Mowing from 1982 to 1993
*Peter W. Dunwiddie, William A. Patterson III,
James L. Rudnicky, and Robert E. Zaremba*

Contents

Experimental Use of Prescribed Fire for Managing ... 99
Grassland Bird Habitat at Floyd Bennett Field,
Brooklyn, New York
*James L. Rudnicky, William A. Patterson III,
and Robert P. Cook*

History of Grasslands in the Northeastern ... 119
United States: Implications for Bird Conservation
Robert A. Askins

Effects of Habitat Area on the ... 137
Distribution of Grassland Birds in Maine
*Peter D. Vickery, Malcolm L. Hunter, Jr.,
and Scott M. Melvin*

Population Viability Analysis for Maine ... 153
Grasshopper Sparrows
Jeffrey V. Wells

Use of Public Grazing Lands by Henslow's ... 171
Sparrows, Grasshopper Sparrows, and Associated
Grassland Birds in Central New York State
Charles R. Smith

Distribution and Population Status of ... 187
Grassland Birds in Massachusetts
Andrea L. Jones and Peter D. Vickery

Grassland Birds in Vermont: Population Status, ... 201
Conservation Problems, and Research Needs
*Charles Darmstadt, Christopher Rimmer,
Judith Peterson, and Christopher Fichtel*

Grassland Bird Habitat Restoration at Floyd Bennett ... 211
Field, Brooklyn, New York: Research and Management
*Richard A. Lent, Thomas S. Litwin,
Robert P. Cook, Jean Bourque,
Ronald Bourque, and John T. Tanacredi*

Lepidopteran Assemblages and the ... 217
Management of Sandplain Communities on
Martha's Vineyard, Massachusetts
Paul Z. Goldstein

Contents

Invertebrate Response to Insecticide Use: 237
What Happens after the Spraying Stops?
*Peter D. Vickery, Jeffrey V. Wells,
and E. Richard Hoebeke*

A Preliminary List of the Insects 251
of the Kennebunk Plains, Maine
*Jeffrey V. Wells, Richard G. Dearborn,
Donald F. Mairs, E. Richard Hoebeke,
Nancy Sferra, Eric C. Roux, Peter D. Vickery,
and Michael A. Roberts*

Status Update and Life History Studies on the 261
Regal Fritillary (Lepidoptera: Nymphalidae)
*David L. Wagner, Matthew S. Wallace,
George H. Boettner, and Joseph S. Elkinton*

X

Contributors

Robert A. Askins is a professor in the Department of Zoology, Connecticut College, New London, CT.

George H. Boettner is a research associate in the Entomology Department, University of Massachusetts, Amherst, MA.

Jean Bourque is a member of the New York City Audubon Society, New York, NY.

Ronald Bourque is a member of the New York City Audubon Society, New York, NY.

Robert P. Cook, previously a National Park Service biologist in Brooklyn, NY, is now a National Park Service biologist in American Samoa.

Charles Darmstadt is director of the North Branch Nature Center, Vermont Institute of Natural Science, Montpelier, VT.

Richard G. Dearborn is an entomologist for the Maine Forest Service, Department of Conservation, Augusta, ME.

Ann C. Dieffenbacher-Krall is a graduate student in the Department of Plant Biology and Pathology, University of Maine, Orono, ME.

Peter W. Dunwiddie, previously the plant ecologist for the Center for Biological Conservation, Massachusetts Audubon Society, Lincoln, MA, is presently a stewardship ecologist for the Washington Field Office of The Nature Conservancy, Seattle, WA.

Joseph S. Elkinton is a professor in the Entomology Department, University of Massachusetts, Amherst, MA.

Christopher Fichtel, formerly of the Nongame and Natural Heritage Program, Vermont Department of Fish and Wildlife, is presently director of science and stewardship, Vermont Field Office, The Nature Conservancy, Montpelier, VT.

Paul Z. Goldstein is a graduate student in the Department of Ecology and Evolutionary Biology, University of Connecticut, Storrs, CT.

E. Richard Hoebeke is assistant curator of the entomology collection, Cornell University, Ithaca, NY.

Malcolm L. Hunter, Jr., is Libra Professor of Conservation Biology, Wildlife Department, University of Maine, Orono, ME.

Andrea L. Jones is the grassland conservation coordinator for the Center for Biological Conservation, Massachusetts Audubon Society, Lincoln, MA, and is a graduate student in the Department of Forestry and Wildlife Conservation, University of Massachusetts, Amherst, MA.

Richard A. Lent is an ecologist and data manager with the Harvard Forest, Petersham, MA.

Thomas S. Litwin is director of the Clark Science Center, Smith College, Northampton, MA.

Contributors

Donald F. Mairs is an entomologist with the Maine Department of Agriculture, Augusta, ME.

Leslie J. Mehrhoff, previously with the Connecticut Geological and Natural History Survey, Connecticut Department of Environmental Protection, is presently curator of the George Safford Torrey Herbarium, Department of Ecology and Evolutionary Biology, University of Connecticut, Storrs, CT.

Scott M. Melvin is the endangered species zoologist for the Natural Heritage and Endangered Species Program, Massachusetts Division of Fisheries and Wildlife, Westborough, MA, and is an adjunct assistant professor in the Department of Forestry and Wildlife Conservation, University of Massachusetts, Amherst, MA.

William A. Patterson III is a professor in the Department of Forestry and Wildlife Conservation, University of Massachusetts, Amherst, MA.

Judith Peterson conducts grassland bird censuses for the Vermont Institute of Natural Science, Woodstock, VT, and lives in Middlebury, VT.

Christopher Rimmer is director of research at the Vermont Institute of Natural Science, Woodstock, VT.

William H. Rivers is chief management forester for the Massachusetts Department of Environmental Management, Amherst, MA.

Michael A. Roberts lives in Steuben, ME.

Eric C. Roux is a graduate student in the Center for Vertebrate Studies, Department of Biology, Northeastern University, Boston, MA.

James L. Rudnicky, previously a research associate in the Department of Forestry and Wildlife Conservation, University of Massachusetts, Amherst, lives in Leverett, MA.

Nancy Sferra is the southern Maine preserve manager for the Maine Chapter of The Nature Conservancy, Sanford, ME.

Charles R. Smith is director of the Arnot Teaching and Research Forest and is a senior research associate, Department of Natural Resources, Cornell University, Ithaca, NY.

John T. Tanacredi is a biologist with the National Park Service, Brooklyn, NY.

Wesley N. Tiffney, Jr., is director of the University of Massachusetts, Boston, Nantucket Field Station, Nantucket, MA.

Peter D. Vickery is avian ecologist for the Center for Biological Conservation, Massachusetts Audubon Society, Lincoln, MA, and is an adjunct assistant professor in the Department of Forestry and Wildlife Conservation, University of Massachusetts, Amherst, MA.

David L. Wagner is an associate professor in the Department of Ecology and Evolutionary Biology, University of Connecticut, Storrs, CT.

Matthew S. Wallace is a graduate student in the Entomology Department, North Carolina State University, Raleigh, NC.

Jeffrey V. Wells, formerly a research associate with the Bird Population Studies Program, Cornell Laboratory of Ornithology, is now coordinator for the New York Important Bird Areas Program, National Audubon Society, Cornell Laboratory of Ornithology, Ithaca, NY.

J. Chris Winne is a research scientist with the Wildlife Spatial Analysis Laboratory at the Montana Cooperative Wildlife Research Unit, University of Montana, Missoula, MT.

Robert E. Zaremba is director of science programs with the New York Regional Office of The Nature Conservancy, Albany, NY.

Introduction

Peter D. Vickery and Peter W. Dunwiddie

Across the landscape of northeastern North America, one sees pervasive evidence of the extensive efforts by European settlers to clear the forest for agriculture. The thousands of kilometers of lichen- and moss-covered stone walls that crisscross the forests of southern and central New England demonstrate the true scope of European settlement, and provide stark evidence of the enormous difficulties these pioneers faced as they opened the land for pastures, hayfields, and croplands.

By the mid-nineteenth century, less than 40 percent of Connecticut, Rhode Island, Massachusetts, and Vermont and only 50 percent of New Hampshire were forested as compared to Maine, which was approximately 80 percent forested (Harper 1918, Powell and Dickson 1984, Frieswyk and Malley 1985a, b, Litvaitis 1993, Whitney 1994). During the nineteenth and early twentieth centuries, however, agriculture declined in the Northeast, even as it expanded in the Midwest. Once again, forests—both hardwood and softwood—reclaimed much of the northeastern North American landscape (Whitney 1994).

Thus, it would be logical to assume that the upland grasslands and heathlands of northeastern North America that are apparent today were the result of European settlement. But was this the case? Did "early-successional" habitats exist in pre-Colonial times? And if they did, how extensive were they? Have these grassland habitats persisted? And if so, what ecological forces have maintained them? Was fire important? Did herbivory by native grazers play an important role? How do land managers today maintain or recreate these early-successional habitats?

How does one reconcile the presence of grassland taxa endemic to northeastern North America with the notion that this region was uninterrupted primal forest? What habitats were available that provided the landscape for the Heath Hen (*Tympanuchus cupido cupido*), an endemic subspecies of the Greater Prairie-Chicken, and several plants and insects endemic to grasslands and early-successional habitats to evolve into unique taxa? Such evolutionary processes often take thousands of years. How could these grassland-shrubland species have survived and evolved without persistent grassland habitat?

What is the present status of grassland birds and insects in New England? What are the prospects of reintroducing the regal fritillary (*Speyeria idalia*), a beautiful grassland butterfly that has completely disappeared from New England since the mid-1970s?

These are some of the questions and ideas that were generated at a conference on native and agricultural grasslands and heathlands of northeastern North America, held at the University of Massachusetts in April 1994. This book is an effort to present recent research—much of which was presented at this conference—on grassland habitats in northeastern North America.

A conference with a specific subject area should bring together much of what is known about that subject. It should inform participants about what is known, but probably more importantly, also reveal what is not known about the subject. Conferences frequently reveal clear geographic biases, and this one had a decided New England emphasis. There were, unfortunately, no papers from the maritime provinces of Canada.

Definition of Grasslands and Heathlands

Grasslands and heathlands encompass a wide variety of types, or "natural communities," from dry sandplain heathlands to bogs, fens, and grass-dominated marshes. At least 13 different types of upland shrubland and herbaceous vegetation "alliances" are described for northeastern North America in a classification developed by The Nature Conservancy (Sneddon et al. 1994). In this book we use the term "native grassland" to encompass both grassland and upland heathland habitats; included within this context are sandplain grasslands, grassland-barrens (commonly called blueberry barrens) of eastern Maine, and ericaceous upland heathlands. Dune grasslands and mountain balds are also included in this definition, although no manuscripts in this volume address these habitat types. Wetland habitats are not included in this definition. We recognize, however, that some "grassland" plants and animals are not restricted to xeric or mesic habitats, and there is one contribution on wetland bluejoint reedgrass (*Calamagrostis canadensis*) grasslands in this volume. In addition, we distinguish native grasslands from agricultural grasslands; the former

Introduction

are dominated by native plants with relatively few introduced or exotic plants. Agricultural grasslands, originally the result of European settlement, are maintained by more recent (i.e., within the past several decades) agricultural activity. These grasslands are intentionally planted with exotic grasses and forbs, such as timothy (*Phleum pratense*) or red clover (*Trifolium pratense*), to establish high-quality forage and hay for farm stock.

Historical Evidence of Grasslands and Heathlands in Northeastern North America

In 1524 the Italian explorer Giovanni da Verrazano entered what is now Narragansett Bay, RI, and found extensive open, treeless plains that stretched for more than 40 kilometers (km)(Day 1953). The Hempstead Plains, a sandplain grassland that once covered more than 24,000 hectares (ha) on Long Island, NY, apparently persisted as a sandplain grassland for several hundred years before it was converted into tract housing, now known as Levittown (Stalter and Lamont 1987; see Askins 1997). Other early European explorers, including Samuel de Champlain and John Smith, also described open areas along the New England coast long before Europeans settled the area (Whitney 1994). And farther east, in Washington County, ME, there is clear evidence that the "pine grassland barrens" of Pineo Ridge have existed as an open grassland-pine (*Pinus*) barren for at least the past 900 years (Winne 1997).

In 1792 two land surveyors described the Epping Plain, near Pineo Ridge, Washington Co., ME: "The plain is the most remarkable in the County, it is capable of parading and maneuvering 40,000 Men" (Pierpont and Albee 1792). Another land surveyor described the site in 1796 as "a plain two or three miles in diameter very poor and barren. The soil is perfectly barren and covered with a short kind of heath and no wood. It has the appearance of having been burned....The nature of the whole is singular and different from anything I ever saw" (A. Baring *in* Fischer 1954).

Although it is prudent to view such historical data with caution (Russell 1983), these early records and observations nevertheless suggest that native grassland habitat existed in patches that were sometimes large, at least occasionally comprising thousands of hectares, for several hundred years before European settlement.

Research into the paleoecological plant record of northeastern grasslands has been especially illuminating. Analyses of pollen cores taken from the bottoms of ponds and wetlands make it possible to reconstruct a general history of the vegetation that dominated the surrounding landscape. Using this technique, Winne (1997) presents strong evidence in this volume for the persistence of a pyrogenically mediated pine-grassland barrens in eastern Maine for at least the past 900 years. However, the origins of these early-successional habitats can differ from region to region. In similar studies, researchers found that the grasslands and heathlands on Nantucket Island and Martha's Vineyard, MA, resulted primarily from European settlement and clearing (Dunwiddie 1989, Stevens 1996).

Also in this volume, Mehrhoff (1997) suggests that many rare grassland plants emigrated to coastal northeastern North America when the "prairie peninsula" extended from the Midwest into Ohio and Pennsylvania during the hypsithermal, a dry period that existed approximately 8,500 to 5,000 years ago. This theory helps explain the occurrence of many plants of midwestern origins or affinities, such as northern blazing star (*Liatris scariosa* var. *novae-angliae*), presently found on the coastal plain of northeastern North America.

Although the paleoecological record is not as clear for animals, particularly birds, Askins (1997) presents compelling evidence for the presence of grassland birds in northeastern North America many years prior to European settlement. It seems unthinkable now that the Heath Hen was once so abundant that laborers and servants balked at eating it more than a few times a week (Nuttall 1840). This bird's former abundance provides a stark reminder of how rapidly overharvesting and habitat destruction can decimate bird populations.

Importance of Fire in Maintaining Early-successional Habitats in Northeastern North America

Fire has helped shape the landscape of northeastern North America since at least the last ice age (Patterson and Sassaman 1988). But examination of the origins of forest fires in this region has revealed that only a small fraction were caused by lightning strikes (Pyne 1984). Most involved human activities (Pyne 1984), and this was probably also the case during the period before European settlement (Patterson and Sassaman 1988). Fires set intentionally by Native Americans were more common in south-coastal New England than in northern or interior New England. From southern Maine along the Saco River and south throughout New England and the Middle Atlantic states, frequent burning for agricultural and hunting purposes has probably occurred for the past three millennia (Patterson and Sassaman 1988).

Because most fires were set by Native Americans, the spatial scale at which fire influenced the landscape was largely shaped by the agricultural and hunting needs of the local Native American population (Cronon 1983). For example, fires set close to permanent coastal villages might have been primarily for agricultural purposes. These sites would have been burned more frequently than those in more remote interior areas or along travel corridors such as rivers, where fires were set primarily for hunting purposes. These fires probably occurred less frequently but likely burned more extensive areas (Patterson and Sassaman 1988).

Frequency of ignition was also affected by the types of fuel available. Fires burn more readily when there is sufficient dry fuel, especially fine fuels, to burn and when weather conditions are favorable for fire to spread (Patterson and Sassaman 1988). Shrubby, ericaceous vegetation, dominant along much of the xeric sandy coastal plain, was probably easier to burn than the shrubs and grasses found in moister habitats. By promoting the spread of fire-tolerant graminoids and other plants whose thin leaves and stems become fine fuels, frequent fires also promote a vegetation type that is more likely to burn in the event of subsequent fires.

Evolutionary Consequences of Fire in Northeastern North America

Fire has been a profoundly important ecological factor for many thousands, if not millions, of years and has strongly affected how numerous plants and animals have evolved. Many plants and animals have adapted to fire, and some species, such as plants with serotinous cones, require fire for their persistence. Conversely, other species do not tolerate fire. For example, maples (*Acer*) and spruces (*Picea*) are usually killed by fire and are absent from areas where fire occurs with any frequency.

In northeastern North America, essentially all native grassland plants are adapted to fire. Some plants, such as little bluestem (*Schizachyrium scoparium*), have extensive fibrous roots to store nutrients. Others, such as northern blazing star and butterfly weed (*Asclepias tuberosa*), have developed large corms or bulbs for the same purpose. Furthermore, it is clear that some species of grassland plants benefit from fire as a form of ecological disturbance. In the case of northern blazing star, for example, fire stimulates flowering, reduces seed predation, and provides a suitable substrate for seed germination (Vickery et al. 1996).

Several grassland birds in northeastern North America have evolved into endemic subspecies or varieties. The Heath Hen, which occurred on grasslands and oak (*Quercus*) barrens only along the eastern seaboard, probably dwelled as far north as southern Maine and south as Maryland and Virginia (Gross 1932). The eastern subspecies of

Henslow's Sparrow (*Ammodramus henslowii susurrans*), which may be extirpated now, had a restricted breeding range from eastern Massachusetts and central New York south to North Carolina (Smith 1968). In maritime Canada, the "Ipswich" Savannah Sparrow's (*Passerculus sandwichensis princeps*) breeding range is restricted to Sable Island, NS (Dwight 1895).

Endemic plants of northeastern grasslands have specific habitat requirements and limited distributions. These plants include bushy rockrose (*Helianthemum dumosum*), sandplain agalinis (*Agalinis acuta*), sickle-leaved golden aster (*Pityopsis falcata* [= *Chrysopsis falcata*]), northern blazing star, and several early-successional shrubs such as shadbush (*Amelanchier*; Wiegand 1912, Jones 1946). Although some grassland plants and animals probably survived in spruce parklands as the Wisconsin glacier receded 18,000 to 12,000 years ago (Guildey et al. 1964, Webb 1988), it is likely that some of these birds and plants colonized northeastern North America and evolved into distinct taxa over the past 5,000 to 8,000 years.

It is notable that many of these endemic taxa are either extinct or now have such restricted ranges that they are threatened with extinction. The Heath Hen is gone, and it is uncertain whether the eastern subspecies of Henslow's Sparrow still exists. Many of the endemic plants and insects are rare and regionally threatened.

European Settlement: Changes in Grassland and Heathland Types

European settlement brought profound changes to the landscapes of northeastern North America. In New England it generally took more than a century before agriculture opened the forest, but by the late eighteenth and early nineteenth centuries more than 60 percent of the forests in Rhode Island, Connecticut, and Massachusetts had been cleared for cropland and pastures (Cronon 1983, Whitney 1994). The expansion of grassland habitat over much of northeastern North America allowed many species, such as the Horned Lark (*Eremophila alpestris*), to expand their ranges and populations (Askins 1993).

Much of the new "grassland" habitat comprised exotic grasses and forbs. Rhizomatous grasses, such as timothy, form a thick, continuous carpet of grass that is generally thicker than native vegetation. Some grassland generalists, such as Savannah Sparrows (*Passerculus sandwichensis*) and Bobolinks (*Dolichonyx oryzivorus*), thrived in this new habitat, but species with more specialized habitat requirements, such as Henslow's Sparrows and Vesper Sparrows (*Pooecetes gramineus*), occupied only specific components of it. Some grassland plants, such as sandplain agalinis, never adapted to this new habitat.

Introduction

Maritime heathlands became extensive in parts of south-coastal New England as early settlers cleared forests from the sandy, droughty morainal and outwash soils along the coast. Introduced pasture grasses did not fare well in these acidic, low-nutrient soils, and many of the native heathland species expanded into these newly opened habitats. Grazing livestock and periodic fires, often ignited by humans, kept the forests from becoming reestablished for several centuries (Dunwiddie 1989). Even after most grazing animals were removed in the late 1800s, heathland plants managed to persist in many areas as salt spray and winds slowed the growth of trees in the open plains of Cape Cod and on Nantucket, Martha's Vineyard, and other coastal islands (Dunwiddie 1989).

Conservation

It is very difficult, if not impossible, to determine how present-day populations of grassland-dependent species compare to those that existed before European settlement. Indeed, given the profound changes that have occurred over the past 400 years to the northeastern North American landscape, it is questionable whether one should view the pre-Colonial landscape as an applicable measure of ecological or biotic integrity for the twentieth and twenty-first centuries. Yet given the abundance of Heath Hens in New England, it seems likely that an array of early-successional habitats supported a wide variety of grassland taxa as well. At a minimum, it is necessary to provide sufficient habitat so that grassland-dependent taxa can persist.

The extent of native and agricultural grasslands in northeastern North America has declined markedly in the past 150 years (Whitney 1994). In the past 60 years the number of hectares of hayfields and pastures in New England and New York has declined by approximately 60 percent (Vickery et al. 1994). As a result, remaining grasslands are more fragmented and isolated. They are also much smaller. In Maine, for example, less than 5 percent of the grassland patches are as large as 32 ha (Vickery et al. 1994). These fragments are often too small to support nesting grassland birds (Vickery et al. 1994) and may not be able to support viable populations of other grassland specialists, especially species that have limited dispersal capabilities such as some butterflies and plants. Not surprisingly, municipal and federal military airports contain some of the largest remaining grassland tracts in southern New England. As a result, these airports now represent the primary refugia in southern New England for area-sensitive birds such as Upland Sandpipers (*Bartramia longicauda*) and Grasshopper Sparrows (*Ammodramus savannarum*; Vickery et al. 1994, Jones and Vickery 1995).

The declines in grassland bird populations in eastern North America and New England since 1960 have been well documented (Vickery 1992, Askins 1993), and it is clear that several grassland species, such as Upland Sandpiper and Grasshopper Sparrow, now breed in the Northeast at only a few isolated sites (Jones and Vickery 1995, Darmstadt et al. 1997). Because breeding populations for several species are small and spatially isolated, using new models, such as Population Viability Analysis, to determine probabilities of long-term persistence is critical for effective conservation planning and action. In this volume, for example, Wells's (1997) analysis of Grasshopper Sparrow viability in Maine shows only a 50 percent likelihood of persistence for 50 years, underscoring the need for this kind of research and for additional conservation action in Maine.

Native grasslands have been reduced to remnant tracts on Cape Cod, Nantucket, Martha's Vineyard, and the Elizabeth Islands, MA, and two isolated sites in southern Maine. On Long Island, NY, only two small patches, 8 and 28 ha, respectively, of the Hempstead Plains (formerly 24,000 ha) remain in a somewhat natural condition (Stalter and Lamont 1987). Extensive native grassland-barren habitat persists in eastern Maine, but because these sites are sprayed regularly with herbicides and insecticides to promote commercial lowbush blueberry (*Vaccinium angustifolium*) production, they are now badly degraded.

Increased isolation and the reduced size of native grassland patches leave few opportunities for managing and conserving these sites on a landscape level. Since most tracts in conservation ownership are relatively small (typically comprising less than 200 ha), it is not possible to manage them in a broad mosaic of early-successional stages. Because many of these sites support rare taxa, they are intensively managed to protect and enhance the rare and threatened species found on them. At the two remaining fragments of the Hempstead Plains, for example, encroachment of weedy species and exotics is a major concern. No more than 60 percent of either site is considered high-quality grassland (R. Zaremba pers. comm.), yet each one harbors between six and nine rare plants and is managed to maintain these endangered species.

Increased isolation and reduced habitat size may have important evolutionary implications for rare grassland species likely to experience genetic bottlenecks and genetic drift (e.g., Schonewald-Cox et al. 1983, Falk and Holsinger 1991). But conservation management intended to protect and enhance rare species is likely to have more profound impacts on the evolutionary courses of rare grassland species than are genetic bottlenecks or drift. For instance, persistent early-spring fires or mowing will ultimately have different evolutionary consequences than will summer burns. Although the impact of a

Introduction

particular management prescription on overall plant composition and abundance has been well documented (Dunwiddie et al. 1997, Rudnicky et al. 1997), little consideration has been given yet to the evolutionary consequences of such management on individual plant and animal species.

Research Needs

In striking contrast to the situation with plants and birds, remarkably little is known about the current status of grassland insects and spiders. Goldstein (1997) makes a convincing case in this volume for the importance of insects as an appropriate, indeed essential, measure of ecological health for grassland and heathland ecosystems. Yet little is known about which insects occupy grassland habitats in northeastern North America, and almost nothing is known about how the insects interact with the flora of these ecosystems. The extent of this ignorance is confirmed in these pages by a preliminary list of invertebrates developed for the Kennebunk Plains, York Co., ME, where at least 10 species are thought to be new state records (Wells et al. 1997). How many new species will be found when the inventory is complete?

The extinction of the regal fritillary from the Northeast further demonstrates our lack of knowledge concerning invertebrates. Although nectar plants for adult regal fritillaries, and food plants for early instars, have been known for many years, essentially nothing is known about the ecological interaction between this butterfly and its host plants. Despite the fact that substantial grassland habitat remains in New England, the regal fritillary has disappeared, and we still do not understand why. Equally disconcerting, it appears that several other species of grassland fritillaries are also declining (Goldstein 1997).

Indeed, for grassland-dependent invertebrates the paucity of information about their distributions, autecology, and interactions with plants and other animals remains one of the largest obstacles to understanding and competently managing native grassland habitats in northeastern North America. The biology of many of the less conspicuous grassland insects is entirely unknown. The distribution of insects and the role they play in grassland ecosystems is an area of inquiry in clear need of careful, quantitative research.

Because many grassland-dependent birds, insects, and plants persist as small populations at isolated sites, it is important to determine whether these populations are viable for the long term. Techniques such as Population Viability Analysis and metapopulation analysis can at least approximate the minimum population levels needed for long-term persistence. Yet to be meaningful, these models require specific demographic and dispersal data that are usually not available. Any studies addressing dispersal capabilities of insects, seeds, or birds would be of paramount value.

Metapopulation analysis can also provide important insights about the kinds of landscape configurations required to maintain viable plant and animal populations. For instance, are habitat corridors important to interchange between isolated patches, and if so, for which species? Viability studies should also include genetic analysis, especially for endemic plants. For example, the two remaining native grasslands on Long Island, NY, provide important opportunities to learn about the genetic viability of small, remnant plant populations. It would be valuable to conduct similar genetic analyses of the possible consequences of isolation and small population sizes for many grassland-dependent taxa in northeastern North America.

Finally, it is imperative that we have a better understanding of the evolutionary consequences of management practices on the taxa being managed. This will require a better understanding of the ecological processes that shaped these grassland habitats and taxa than we have today. Careful, quantitative research framed around present management practices on grassland habitats provides excellent opportunities to learn more about these grassland ecosystems, and to incorporate these results into improved management practices.

Acknowledgments

R. Askins, C. Duncan, C. Griffin, J. Herkert, W. A. Patterson III, W. G. Shriver, and B. Vickery reviewed this manuscript and made numerous suggestions that improved it greatly. Thank you. Critical logistical support was provided by E. Bell, J. Bodge, S. Vickery, G. Vickery, and B. Vickery. We especially thank C. Griffin. His support and encouragement were invaluable and much appreciated.

Literature Cited

Askins, R. A. 1993. Population trends in grassland, shrubland, and forest birds in eastern North America. Current Ornithol. 11: 1–34.

Askins, R. A. 1997. History of grasslands in the northeastern United States: implications for bird conservation. Pp. 119–136 *in* Grasslands of northeastern North America: ecology and conservation of native and agricultural landscapes (P. D. Vickery and P. W. Dunwiddie, eds.). Massachusetts Audubon Soc., Lincoln.

Cronon, W. 1983. Changes in the land: Indians, colonists, and the ecology of New England. Hill and Wang, New York.

Darmstadt, C., C. Rimmer, J. Peterson, and C. Fichtel. 1997. Grassland birds in Vermont: population status, conservation problems, and research needs. Pp. 201–210 *in*

Introduction

Grasslands of northeastern North America: ecology and conservation of native and agricultural landscapes (P. D. Vickery and P. W. Dunwiddie, eds.). Massachusetts Audubon Soc., Lincoln.

Day, G. M. 1953. The Indian as an ecological factor in the northeast forest. Ecology 34: 329-346.

Dunwiddie, P. W. 1989. Forest and heath: the shaping of vegetation on Nantucket Island. J. Forest Hist. 33: 126-133.

Dunwiddie, P. W., W. A. Patterson III, J. L. Rudnicky, and R. E. Zaremba. 1997. Vegetation management in coastal grasslands on Nantucket Island, Massachusetts: effects of burning and mowing from 1982 to 1993. Pp. 85-98 *in* Grasslands of northeastern North America: ecology and conservation of native and agricultural landscapes (P. D. Vickery and P. W. Dunwiddie, eds.). Massachusetts Audubon Soc., Lincoln.

Dwight, J., Jr. 1895. The Ipswich Sparrow (*Ammodramus princeps* Maynard) and its summer home. Mem. Nuttall Ornithol. Club no. 2, Cambridge, MA.

Falk, D. A., and K. E. Holsinger. 1991. Genetics and conservation of rare plants. Oxford Univ. Press, Oxford, U.K.

Fischer, R. A. 1954. Calendar of the letters of Alexander Baring, 1795-1801. Manuscript Div., Library of Congress, Washington, D.C.

Frieswyk, T. S., and A. M. Malley. 1985a. Forest statistics for Vermont: 1973 and 1983. Research Bull. NE-87, U.S. Forest Serv., Northeast Forest Exper. Stn., Broomall, PA.

Frieswyk, T. S., and A. M. Malley. 1985b. Forest statistics for New Hampshire: 1973 and 1983. Research Bull. NE-88, U.S. Forest Serv., Northeast Forest Exper. Stn., Broomall, PA.

Goldstein, P. Z. 1997. Lepidopteran assemblages and the management of sandplain communities on Martha's Vineyard, Massachusetts. Pp. 217-236 *in* Grasslands of northeastern North America: ecology and conservation of native and agricultural landscapes (P. D. Vickery and P. W. Dunwiddie, eds.). Massachusetts Audubon Soc., Lincoln.

Gross, A. O. 1932. Heath Hen. Pp. 264-280 *in* Life histories of North American gallinaceous birds (A. C. Bent, ed.). U.S. Natl. Mus. Bull. 162.

Guildey, J. E., P. S. Martin, and A. D. McGrady. 1964. New Paris no. 4: a Pleistocene cave deposit in Bedford County, Pennsylvania. Bull. Natl. Speleolog. Soc. 26: 121-194.

Harper, R. M. 1918. Changes in forest area of new England in three centuries. J. Forestry 16: 442-452.

Jones, A. L., and P. D. Vickery. 1995. Distribution and population status of grassland birds in Massachusetts. Bird Observer 23: 89-96.

Jones, G. N. 1946. American species of *Amelanchier*. Ill. Biol. Monogr. 20.

Litvaitis, J. A. 1993. Response of early successional vertebrates to historic changes in land use. Conserv. Biol. 7: 866-873.

Mehrhoff, L. J. 1997. Thoughts on the biogeography of grassland plants in New England. Pp. 15-24 *in* Grasslands of northeastern North America: ecology and conservation of native and agricultural landscapes (P. D. Vickery and P. W. Dunwiddie, eds.). Massachusetts Audubon Soc., Lincoln.

Nuttall, T. 1840. A manual of ornithology. 2d ed. Hilliard, Gray and Co., Boston.

Patterson, W. A., III, and K. E. Sassaman. 1988. Indian fires in the prehistory of New England. Pp. 107-135 *in* Holocene human ecology in northeastern North America (G. P. Nichols, ed.). Plenum, New York.

Pierpont, J., and W. Albee. 1792. A journal over 1,000000 acres of land in the counties of Hancock and Washington. Handwritten journal. Historical Soc., Philadelphia, PA.

Powell, D. S., and D. R. Dickson. 1984. Forest statistics for Maine: 1971 and 1982. Research Bull. NE-81, U.S. Forest Serv., Northeast Forest Exper. Stn., Broomall, PA.

Pyne, S. J. 1984. Introduction to wildland fire. John Wiley, New York.

Rudnicky, J. L., W. A. Patterson III, and R. P. Cook. Experimental use of prescribed fire for managing grassland bird habitat at Floyd Bennett Field, Brooklyn, New York. Pp. 99-118 *in* Grasslands of northeastern North America: ecology and conservation of native and agricultural landscapes (P. D. Vickery and P. W. Dunwiddie, eds.). Massachusetts Audubon Soc., Lincoln.

Russell, E. W. B. 1983. Indian-set fires in the forests of the northeastern United States. Ecology 64: 78-88.

Schonewald-Cox, C. M., S. M. Chambers, B. MacBryde, and W. L. Thomas. 1983. Genetics and conservation. Benjamin/Cummings Publ., Menlo Park, CA.

Smith, W. P. 1968. Eastern Henslow's Sparrow. Pp. 776-778 *in* Life histories of North American cardinals, grosbeaks, buntings, towhees, finches, sparrows, and allies (A. C. Bent, ed.). U.S. Natl. Mus. Bull. 237.

Sneddon, L., M. Anderson, and K. Metzler. 1994. A classification and description of terrestrial community alliances in The Nature Conservancy's eastern region. The Nature Conservancy, Boston.

Stalter, R., and E. E. Lamont. 1987. Vegetation of the Hempstead Plains, Mitchell Field, Long Island, New York. Bull. Torrey Bot. Club 114: 330-335.

Stevens, A. 1996. The paleoecology of coastal sandplain grasslands on Martha's Vineyard, Masssachusetts. Ph.D. diss., Univ. of Massachusetts, Amherst.

Vickery, P. D. 1992. A regional analysis of endangered, threatened and special concern birds in the northeastern United States. Trans. N. E. Sect. Wildl. Soc. 48: 1-10.

Vickery, P. D., T. Enz, and P. W. Dunwiddie. 1996. Influence of fire on the reproductive ecology of northern blazing star, a rare grassland perennial of northeastern North America. Bull. Ecol. Soc. Am. 77: 458.

Vickery, P. D., M. L. Hunter, Jr., and S. M. Melvin. 1994. Effects of habitat area on the distribution of grassland birds in Maine. Conserv. Biol. 8: 1087-1097.

Webb, T., III. 1988. Eastern North America. Pp. 385-414 *in* Vegetation history (B. Huntley and T. Webb III, eds.). Kluwer Academic Publ., Hingham, MA.

Wells, J. W. 1997. Population viability analysis for Maine Grasshopper Sparrows. Pp. 153-170 *in* Grasslands of northeastern North America: ecology and conservation of native and agricultural landscapes (P. D. Vickery and P. W. Dunwiddie, eds.). Massachusetts Audubon Soc., Lincoln.

Wells, J. V., R. G. Dearborn, D. F. Mairs, E. R. Hoebeke, N. Sferra, E. C. Roux, P. D. Vickery, and M. A. Roberts. 1997. A preliminary list of the insects of the Kennebunk Plains, Maine. Pp. 251-260 *in* Grasslands of northeastern North America: ecology and conservation of native and agricultural landscapes (P. D. Vickery and P. W. Dunwiddie, eds.). Massachusetts Audubon Soc., Lincoln.

Whitney, G. G. 1994. From coastal wilderness to fruited plain. Cambridge Univ. Press, Cambridge, U.K.

Wiegand, K. M. 1912. The genus *Amelanchier* in eastern North America. Rhodora 14: 117-161.

Introduction

Winne, J. C. 1997. History of vegetation and fire on the Pineo Ridge pine grassland barrens of Washington County, Maine. Pp. 25-52 *in* Grasslands of northeastern North America: ecology and conservation of native and agricultural landscapes (P. D. Vickery and P. W. Dunwiddie, eds.). Massachusetts Audubon Soc., Lincoln.

Peter D. Vickery: Center for Biological Conservation, Massachusetts Audubon Society, Lincoln, MA 01773, and Department of Forestry and Wildlife Conservation, University of Massachusetts, Amherst, MA 01003.

Peter W. Dunwiddie: Center for Biological Conservation, Massachusetts Audubon Society, Lincoln, MA 01773 (present address: The Nature Conservancy, Washington Field Office, Seattle, WA 98101).

Thoughts on the Biogeography of Grassland Plants in New England

Leslie J. Mehrhoff

Introduction

When most naturalists think of grasslands in North America, they immediately think of midwestern prairies. Not until the modifier "New England" is added do they think of some of the grasslands of the Northeast, such as the Kennebunk Plains, ME, or the grasslands on Martha's Vineyard and Nantucket Island, MA, and on Block Island, RI. Some naturalists consider Connecticut's North Haven Sand Plains (Britton 1903, Olmsted 1937) and the heathlands of Nantucket (Jorgensen 1971) as examples of less typical New England grasslands.

Biogeographically, Long Island, NY, also should be considered part of New England as it was formed by the outwash plains and terminal moraines of glaciers that extended south across New England (Klemens 1993). Long Island Sound is an artifact of its predecessor, glacial Lake Connecticut (Patton and Kent 1992). Long Island's well-known Hempstead Plains (Harper 1911) should undoubtedly be included in a list of "New England" grasslands. Until recently, only ornithologists probably would have included some of the larger regional airports such as Bradley Field north of Hartford, CT, or military bases such as Westover Air Force Base near Springfield, MA, in lists of New England grasslands. These areas appear to be the last survivia for several coastal-plain plant species and for some threatened or endangered grassland birds.

Some of the grasslands mentioned above are reasonably large, especially by New England standards. Many of New England's grasslands have been repeatedly fragmented until only small parcels of grassland remain. Individual pockets of grassland are usually small and frequently go unnoticed. Although not large enough to support viable populations of grassland birds (Vickery et al. 1994, R. Askins pers. comm.), these small grasslands could potentially support populations of many invertebrate species (D. Schweitzer pers. comm., D. Wagner pers. comm.) and their host plants, as well as localized populations of some of New England's rare vascular plants.

Many, if not all, of New England's remaining grasslands have been strongly influenced by humans for many years. Many of these grasslands were created intentionally, and most have a history of heavy human disturbance since Colonial times (Thompson 1977, Russell 1980). Other smaller grasslands, such as open, rocky balds, although natural in the context of the current vegetation, are usually ephemeral.

Many of the native plant species associated with New England's grasslands are widespread and common throughout the region (Table 1). None are entirely restricted to grasslands, and in fact many have spread to roadsides and other artificially maintained but ecologically similar habitats. Other vascular plant species found in New England grasslands are rare in some states or throughout New England (Table 2). Many of these are listed as threatened or endangered in some states, although a few of these rarities may be locally abundant at particular sites.

Biogeography

Biogeographic affinities of New England's grassland flora may shed some light on the origin of grassland systems in the Northeast. All of the common vascular plant species of New England's grasslands are wide ranging and have affinities with the midwestern prairie flora or the southern coastal-plain flora. The questions of how and when these floristic elements arrived in New England, how they dispersed throughout the region, and where they grew in presettlement times are all subject to investigation. The ideas examined in this paper are based primarily on the current distribution of certain grassland taxa; no quantitative analysis has been conducted. Detailed investigations into local paleoecology and paleoclimatology of New England's grasslands are needed to substantiate or refute these ideas.

With the exception of parts of Nantucket and Martha's Vineyard, MA, and Long Island, NY, all of New England was covered by ice during the Wisconsin glaciation (Chamberlain 1981, Bell 1985, Kendall 1987, Oldale 1992, Patton and Kent 1992). As the glaciers retreated, they left newly exposed soil that was colonized by plants expanding their ranges from isolated remnant pockets in the south, west, or exposed coastal shelf off New England. As temperatures warmed and

Biogeography of New England's Grassland Flora

Table 1. *Common native grassland plant species of northeastern North America. Some of these taxa may be rare in certain states but generally are widespread.*

scrub oak (*Quercus ilicifolia*)
scrub chestnut oak (*Quercus prinoides*)
frostweed (*Helianthemum canadense*)
ovate-leaved violet (*Viola fimbriatula*)
birdfoot violet (*Viola pedata*)
round-headed bush-clover (*Lespedeza capitata*)
hairy bush-clover (*Lespedeza hirta*)
racemed milkwort (*Polygala polygama*)
blunt-leaved milkweed (*Asclepias amplexicaulis*)
stiff aster (*Aster linariifolius*)
New England aster (*Aster novae-angliae*)
small flat-topped aster (*Aster paternus*)
heath aster (*Aster pilosus*)
grass-leaved goldenrod (*Euthamia graminifolia*)
tall goldenrod (*Solidago altissima*)
late goldenrod (*Solidago gigantea*)
early goldenrod (*Solidago juncea*)
gray goldenrod (*Solidago nemoralis*)
downy goldenrod (*Solidago puberula*)
rough-stemmed goldenrod (*Solidago rugosa*)
rush (*Juncus greenei*)
sedge (*Carex lucorum* or *Carex pensylvanica*)
big bluestem (*Andropogon gerardii*)
poverty grass (*Danthonia spicata*)
rice grass (*Oryzopsis pungens*)
little bluestem (*Schizachyrium scoparium*)
indian grass (*Sorghastrum nutans*)
fresh-water cord-grass (*Spartina pectinata*)

the ice retreated, plant species began to disperse across the landscape. This does not imply that changes in vegetation would have been visible over the short term but rather that a series of floristic advancements and successions occurred. It is likely that by 10,000 years before present (YBP) most of what is now New England was covered with some type of vegetation (Pielou 1991).

Lists of vascular plant species from midwestern prairies have striking similarities with those from New England (Gleason 1912, 1923, Schaffner 1913, Curtis 1959, King 1981, Madany 1981, Thompson 1981). The presence of many of the same species in both regions indicates that New England grasslands have a floristic affinity with

Table 2. Plants that are rare or restricted on New England grasslands.

bushy rockrose (*Helianthemum dumosum*)
low rockrose (*Helianthemum propinquum*)
wild lupine (*Lupinus perennis*)
sandplain flax (*Linum intercursum*)
butterfly weed (*Asclepias tuberosa*)
sandplain agalinis (*Agalinis acuta*)
eastern silvery aster (*Aster concolor*)
sickle-leaved golden aster (*Pityopsis falcata* [= *Chrysopsis falcata*])
purple cudweed (*Gamochaeta purpurea*)
broad-leaved golden aster (*Chrysopsis mariana*)
northern blazing star (*Liatris scariosa* var. *novae-angliae*)
sandplain blue-eyed grass (*Sisyrinchium arenicola* [= *fuscatum*])

midwestern prairies. These floristic elements may have arrived in the Northeast during a period when regional climates were warmer and drier.

The maximum warm period after the Wisconsin glaciation, known as the hypsithermal or xerothermic period, occurred from about 8,500 to 5,000 YBP (Pielou 1991). This is the period when the "prairie peninsula," first described by Adams (1902) and characterized by Transeau (1935), extended farthest east (Fig. 1).

The typical representation of the eastern extreme of the prairie peninsula looks something like a closed fist, reaching east into present-day Ohio and extreme western Pennsylvania. A more realistic representation might be an open hand with fingers outstretched to the east and northeast. Gleason (1923) described eastern extensions of prairie vegetation as "amoeba-like arms." It was during the xerothermic period that elements of the prairie flora most likely extended eastward into New England.

A possible route for some of these plant migrations may have been through the lowlands of central New York. Based on existing distribution patterns of certain prairie species in the Northeast, it is possible to surmise that the plant species may have migrated from Ohio and western Pennsylvania east through western New York, then along the Mohawk River valley, possibly arriving at the Hudson River valley around Albany, then dispersing to the east and south (Fig. 1). Transeau (1935) wrote that most prairie taxa did not cross the Allegheny Mountains but instead arrived in the Northeast via New York or along the expanded coastal plain from the south. Gordon (1940) also noted that prairie species had migrated eastward into lowland New York. Sears (1942) felt that "while the evidence is complex and not wholly consistent, we have every reason to believe that for North America more than one

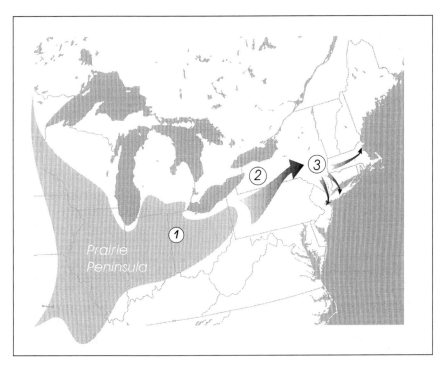

Figure 1. Numeral 1: Prairie peninsula (redrawn from Stuckey 1981); numeral 2: hypothetical eastward dispersal route of midwestern prairie fauna; numeral 3: hypothetical secondary dispersal routes in Northeast along coastal plain.

opportunity has existed for the northeastward movement of vegetation and floristic boundaries."

Whereas some prairie plants repopulated the glaciated Northeast via the Mississippi embayment/coastal-plain route south around the Appalachian Mountains, I suspect that the majority of the prairie elements arrived via the overland route through low-elevation gaps in the Appalachian Mountains. This hypothesis seems to be corroborated for other taxa, especially reptiles and amphibians (Schmidt 1938).

It is well established that Native Americans cleared land for agriculture or wildlife, and this may have duplicated conditions prevalent during the hypsithermal. It has been suggested (J. Slater pers. comm.) that midwestern taxa may have migrated eastward during or shortly after European settlement and expansion. If this were the case, early botanical collectors likely would have documented many of these midwestern grassland plants as being more widely distributed than they are today.

Prairie elements along the coastal plain could have arrived from a secondary dispersal southeastward and eastward, as the hypsithermal ended and New England cooled and became more moist. Species such as stiff goldenrod (*Solidago rigida*), heart-leaved alexanders (*Zizia aptera*), many grasses such as dropseed (*Sporobolus compositus* var. *compositus*), and possibly northern blazing star (*Liatris scariosa* var. *novae-angliae*) may have arrived in the Northeast via this route.

Does the presence in New England of midwestern "prairie" plants mean that at one time prairies existed in New England similar to those occurring today in the Midwest? Probably not. Although Sears (1942) suggested that flora also changed during the xerothermic period, conditions have not remained dry enough or warm enough for prairies to persist in New England. Furthermore, according to Cowles (1928), not enough time has occurred since the last glacial retreat for prairie soils to develop in this region. These factors, however, do not preclude the existence of prairie elements in New England's flora.

Certain other grassland taxa have affinities with the Atlantic coastal plain and appear to be regionally endemic to the Northeast. These taxa probably arrived in the Northeast by migrating from survivia on the coastal plain along an exposed coastal shelf.

No northeastern grassland endemics occur entirely north of the Wisconsin glacial maximum. Some northeastern endemics found in grasslands, such as bushy rockrose (*Helianthemum dumosum*), sickle-leaved golden aster (*Pityopsis falcata* [= *Chrysopsis falcata*]), and sandplain agalinis (*Agalinis acuta*), probably existed on the expanded coastal-plain shelf or south of the glacial maximum on the coastal plain in midcoastal states. These northeastern coastal-plain endemics are biogeographically significant.

Floristic Associations

Where did these grassland elements occur in pre-Colonial times? Coastal-plain taxa such as sandplain flax (*Linum intercursum*), eastern silvery aster (*Aster concolor*), broad-leaved golden aster (*Chrysopsis mariana*), and possibly sickle-leaved golden aster and bushy rockrose probably occurred on New England's coastal islands. Other grassland species, such as sandplain agalinis, wild lupine (*Lupinus perennis*), butterfly weed (*Asclepias tuberosa*), and low rockrose (*Helianthemum propinquum*), probably have always existed in dry, open sandplains or scrub oak (*Quercus* spp.) barrens near the coast. These sites are not necessarily true coastal-plain habitats in the strict geological sense, as much of New England no longer has an exposed coastal plain. Many "barrens" originated as glacial lake bottoms or along glacial river valleys where sands and gravels were deposited after the glaciers withdrew from the region.

The remainder of the grassland taxa possibly existed in a mosaic of small grassy patches such as river floodplains, exposed rocky balds, glacial lake dunes, and aeolian deposits, or in other sites with similar conditions. Some of these openings were natural, but others may have been created intentionally or accidentally by Native Americans. Fires must have been important in maintaining these open areas. New habitats would have been created by fire, drought, or other natural phenomena. Enough of these sites must have existed to allow recruitment of taxa into newly created habitats as older sites were lost to succession.

Conservation

Some of the vascular plant taxa that occur on New England grasslands are rare. These state rarities need some form of protection. In most instances, these vulnerable species can best be preserved by managing the entire grassland system by arresting succession. If a diversity of floristic elements is to be retained in New England, it is imperative that sound management plans for all remaining grasslands be implemented. There is no time to lose.

The presence of rare or endangered vascular plant species can help identify those sites in need of protection and may promote the protection of the entire grassland system (Dunwiddie et al. 1993). Current conservation of New England grasslands must focus on entire systems if it is to be effective.

Summary

Prairie elements presently occur in New England grasslands, but it is unlikely that New England ever had extensive prairie communities. A possible and notable exception is the Hempstead Plains on Long

Island, NY. This does not suggest that small, restricted sites do not, or did not, exist on the New England landscape. In fact, they look remarkably similar to prairies of the Midwest, writ small.

Grassland floristic elements probably arrived by one of two postglacial migratory routes. Some grassland taxa have affinities to midwestern prairies and probably migrated from that region during the xerothermic period. Other taxa, including some regional endemics, arrived from the south, having migrated along an expanded coastal plain. It is also possible that some of these coastal-plain taxa survived glaciation in survivia on the now-submerged coastal shelf off New England.

Regardless of how these rare grassland plants arrived in New England, it is clear that they are in urgent need of protection.

Acknowledgments

These ideas have been discussed with many naturalists, notably S. P. Cover, J. J. Dowhan, R. Enser, C. L. Remington, J. A. Slater, B. A. Sorrie, P. D. Vickery, and most importantly A. W. H. Damman. My appreciation for my colleagues' time does not imply that they concur with the ideas presented here. P. W. Dunwiddie and P. D. Vickery made many helpful comments on an earlier draft of this paper. The maps were prepared by M. J. Spring.

Literature Cited

Adams, C. C. 1902. Postglacial origin and migrations of the life of the northeastern United States. J. Geogr. 1: 352-357.

Bell, M. 1985. The face of Connecticut: people, geology, and the land. State Geol. and Nat. Hist. Surv. Conn. Bull. 110.

Britton, W. E. 1903. Vegetation of the North Haven sand plains. Bull. Torrey Bot. Club. 30: 571-620.

Chamberlain, B. B. 1981. These fragile outposts: a geological look at Cape Cod, Martha's Vineyard, and Nantucket. Parnassus Imprints, Yarmouth Port, MA.

Cowles, H. C. 1928. Persistence of prairies. Ecology 9: 380-382.

Curtis, J. T. 1959. The vegetation of Wisconsin. Univ. of Wisconsin Press, Madison.

Dunwiddie, P. W., K. A. Harper, and R. A. Zaremba. 1993. Classification and ranking of coastal heathlands and sandplain grasslands in Massachusetts. Report to Massachusetts Natural Heritage and Endangered Species Prog., Boston.

Gleason, H. A. 1912. An isolated prairie grove and its phytogeographical significance. Bot. Gaz. 53: 38-49.

Gleason, H. A. 1923. The vegetational history of the middle west. Ann. Assoc. Am. Geog. 12: 39-85.

Gordon, R. B. 1940. The primeval forest types of southwestern New York. N.Y. State Mus. Bull. 321.

Harper, R. M. 1911. The Hempstead plains: a natural prairie on Long Island. Bull. Am. Geog. Soc. 43: 351-360.

Jorgensen, N. 1971. A guide to New England's landscape. Barre Publ. Co., Barre, MA.

Kendall, D. L. 1987. Glaciers and granite: a guide to Maine's landscape and geology. Down East Books, Camden, ME.

King, C. C. 1981. Prairies of the Darby plains in west-central Ohio. Pp. 108-127 *in* The prairie peninsula—in the shadow of Transeau (R. L. Stuckey and K. J. Reese, eds.). Ohio Biol. Surv. Notes no. 15.

Klemens, M. W. 1993. Amphibians and reptiles of Connecticut and adjacent regions. State Geol. and Nat. Hist. Surv. Conn. Bull. 112.

Madany, M. H. 1981. A floristic survey of savannas in Illinois. Pp. 177-181 *in* The prairie peninsula—in the shadow of Transeau (R. L. Stuckey and K. J. Reese, eds.). Ohio Biol. Surv. Notes no. 15.

Oldale, R. N. 1992. Cape Cod and the islands: the geologic story. Parnassus Imprints, Yarmouth Port, MA.

Olmsted, C. E. 1937. Vegetation of certain sand plains of Connecticut. Bot. Gaz. 99: 209-300.

Patton, P. C., and J. M. Kent. 1992. A moveable shore: the fate of the Connecticut coast. Duke Univ. Press, Durham, NC.

Pielou, E. C. 1991. After the ice age: the return of life to glaciated North America. Univ. of Chicago Press, Chicago.

Russell, H. S. 1980. Indian New England before the Mayflower. Univ. Press of New England, Hanover, NH.

Schaffner, J. H. 1913. The characteristic plants of a typical prairie. Ohio Nat. 13: 65-69.

Schmidt, K. P. 1938. Herpetological evidence for the postglacial eastward extension of the steppe in North America. Ecology 19: 396-407.

Sears, P. B. 1942. Xerothermic theory. Bot. Rev. 8: 708-736.

Stuckey, R. L. 1981. Origin and development of the concept of the prairie peninsula. Pp. 4-23 *in* The prairie peninsula—in the shadow of Transeau (R. L. Stuckey and K. J. Reese, eds.). Ohio Biol. Surv. Notes no. 15.

Thompson, B. F. 1977. The changing face of New England. Houghton Mifflin Co., Boston.

Thompson, P. W. 1981. Flora of Dayton prairie, a remnant of Terre Coupee prairie, in Michigan. Pp. 148-150 *in* The prairie peninsula—in the shadow of Transeau (R. L. Stuckey and K. J. Reese, eds.). Ohio Biol. Surv. Notes no. 15.

Transeau, E. N. 1935. The prairie peninsula. Ecology 16: 423-437.

Vickery, P. D., M. L. Hunter, Jr., and S. M. Melvin. 1994. Effects of habitat area on the distribution of grassland birds in Maine. Conserv. Biol. 8: 1087-1097.

Leslie J. Mehrhoff: George Safford Torrey Herbarium, Box U-42, Department of Ecology and Evolutionary Biology, University of Connecticut, Storrs, CT 06269-3042.

History of Vegetation and Fire on the Pineo Ridge Pine Grassland Barrens of Washington County, Maine

J. Chris Winne

Abstract

The history of vegetation and fire on the Pineo Ridge emergent glaciomarine delta in eastern coastal Maine was inferred from the analysis of fossil pollen and microscopic charcoal from the upper sediments of two paired basins, Pineo Pond (on the delta) and Mud Pond (off the delta), as well as from historical records. The study was prompted by the question of whether the current open aspect of the delta vegetation is a "natural" landscape predating European influence.

Pineo Pond is a small kettle lake located on the sandy delta. For more than 100 years, much of the delta has been maintained in an open condition for commercial production of lowbush blueberry (*Vaccinium angustifolium*). Historical accounts indicate that prior to commercial use, some parts of the delta were open whereas others were dominated by pines (*Pinus*). Mud Pond is located 10 kilometers (km) southwest in an area of greater relief and finer textured soils. The modern vegetation around Mud Pond is mixed northern hardwoods, with American beech (*Fagus grandifolia*), birch (*Betula*), spruce (*Picea*), eastern hemlock (*Tsuga canadensis*), maple (*Acer*), and northern red oak (*Quercus rubra*) being common associates.

Sediment analysis indicated that for the period studied (about the past 1,800 years) the vegetation near Pineo Pond has differed distinctly from that near Mud Pond. A consistently more persistent pollen component of species associated with short-return-interval fires, such as white pine (*Pinus strobus*), red pine (*P. resinosa*), sweet fern (*Comptonia peregrina*), and bracken (*Pteridium aquilinum*), was found in the sediments of Pineo Pond than in those of Mud Pond. Charcoal analysis indicated that the area around Pineo Pond had more fires, but of a less catastrophic nature, than the area around Mud Pond. The evidence suggested that a major fire about 1,700 years ago created a grassland pine/shrub barren on the Pineo Ridge delta. This open vegetation has persisted to some extent to the present day.

Introduction

Fire is important in maintaining the structure and integrity of many forested and nonforested ecosystems (Kozlowski and Ahlgren 1974, Wein and MacLean 1983). Its role differs among ecosystems, depending on the particular local and regional relationships between fire interval, fire intensity, and mean fire rotation of the existing vegetation (Heinselman 1981). Where fire frequency is high, ecosystems may depend on fire for their renewal; where fire frequency is lower, however, greater disruption of established vegetation results (White 1979).

Fire-return intervals generally decrease in North America from the midwestern prairie-forest boundary to the East Coast as a result of continental patterns of precipitation, day length, and potential ignition sources (Wein and MacLean 1983). The extensive jack pine (*Pinus banksiana*) forests of the Minnesota Boundary Waters Canoe Area may have a fire rotation of 100 years or less and are dependent on fire for their long-term place in the vegetation mosaic of that area (Heinselman 1973). In most eastern locations, investigators have inferred much longer fire-rotation periods, ranging from 350 years (Green 1981) to about 800 years (Lorimer 1977) and even to more than 1,000 years (Wein and Moore 1977, 1979, Fahey and Reiners 1981). With a fire rotation greater than the life span of the dominant tree species, fire does not provide stability in most eastern forests, but rather may initiate extensive reorganization of species groupings (Green 1982).

On a more local scale, fire can contribute to vegetation pattern. The occurrence of fire varies locally with topography, vegetation, moisture regime, soil type, and occurrence of fire breaks (Foster 1983). Areas of flatter topography and coarser, more xeric soils tend to have shorter return intervals and a more fire-adapted vegetation than do nearby areas with greater relief and finer soils (Zackrisson 1977, Grimm 1984). Fahey and Reiners (1981) have suggested that minimum rotation periods for the pine forests of the coasts of southern Maine and New Hampshire are comparable to those in the boreal coniferous forests.

This paper examines the role of fire in creating and maintaining the unusually open vegetation found on an area known as Pineo Ridge in eastern Maine. The grass/shrubland vegetation and relatively flat topography contrast sharply with surrounding areas. At present the delta "barrens" (as they are known locally) are used for commercial blueberry production and are maintained in an open condition by periodic burning. In fields that are not maintained, fire-dependent grasses, shrubs, and red and white pines often predominate. This paper considers the hypothesis that the open vegetation is a natural landscape that has only been manipulated, not created, by modern human intervention.

Study Region

Pineo Ridge is an emergent glaciomarine delta in western Washington County, ME (Fig. 1). This 45-km^2 (Borns 1979) delta is the largest glaciomarine delta in Maine (Stone 1899) and has sand and gravel deposits up to 70 meters (m) deep and with relatively low surface relief. Vegetationally the delta lies between a northern mixed-hardwood zone to the north and west and a northeastern spruce-fir (*Abies*) boreal strip on the coast to the south and east (Davis 1966, Küchler 1975). The northern mixed hardwoods are an association of American beech, maples, birches, balsam fir (*Abies balsamea*), red spruce (*Picea rubens*), white pine, northern red oak, and eastern hemlock. Spruces and firs mixed with red maple (*Acer rubrum*) and birches prevail in the boreal zone. Peatlands with black spruce (*Picea mariana*) are numerous in the area (Davis et al. 1983).

The delta grassland barrens differ considerably from the surrounding areas. Much of the delta surface has been used for commercial lowbush blueberry production for more than 100 years, and it is maintained in an open condition by burning, mowing, and, since the mid-1980s, treatment with herbicides. Managed areas are strongly dominated today by lowbush blueberry. The shrublands found in areas of less-intensive management contain higher frequencies of highbush blueberry (*Vaccinium corymbosum*), sheep laurel (*Kalmia angustifolia*), sweet fern, bracken, and other fire-adapted species. Where forest vegetation is established, gray birch (*Betula populifolia*), paper birch (*B. papyifera*), red and white pines, and northern red oaks are common associates. Pines and oaks often have multiple fire scars, illustrating their tolerance of recurring fires. Historically the delta was forested with a pine barren/shrubland heath (Pierpont and Albee 1792), a composition that suggests an important long-term role for fire in maintaining the character of the local vegetation.

This study used two complementary tools to address the origin of the delta grasslands. Fossil pollen and charcoal from the sedimentary record of two small ponds (one on and one off the delta) were analyzed and compared, and the historical record was examined to understand the nature of the region from the time of early settlement to the present. The study was concerned with the following questions: (1) Have open shrublands been consistently present on the Pineo Ridge delta for more than the past few centuries following European settlement? (2) Do comparisons of fire history from the Pineo Ridge delta and a nearby site support the hypothesis that localized topography and soils affect fire frequency? (3) Does the present-day open-grown and fire-adapted vegetation of the delta result from the long-term impact of geomorphology and soils? (4) Can differences between the present vegetation of the delta and that of the nearby surrounding area

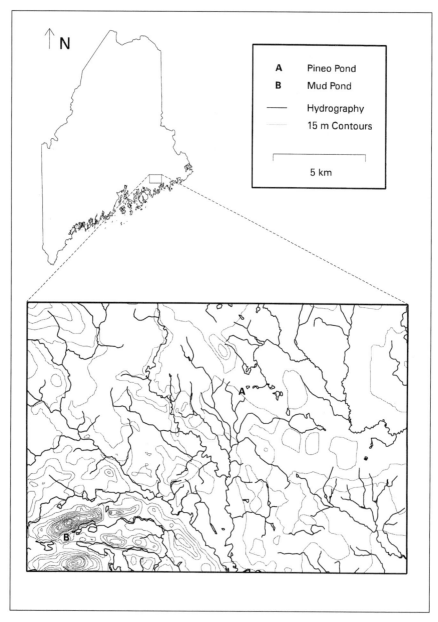

Figure 1. Topography and hydrography of the study region in western Washington County, ME.

be attributed to differing local fire regimes, or are these differences more likely a result of historical cultural influences? and (5) Can natural rates of fire ignition account for the frequency of fires inferred from the sedimentary record?

The overall design of the paleoecological investigation incorporated the "paired basin" technique developed by Jacobson (1979). Palynological data from Pineo Pond, which is located on the sandy, well-drained delta, were compared with those from Mud Pond, which is located nearby in an area of greater local relief and shallow but more mesic soils. Local and regional inputs were separated by subtracting time-stratified data of one site from those of the other. The technique assumed that regional inputs to the sedimentary record were similar in nearby contrasting small basins of similar morphology, and that differences in the data could be attributed to local differences in vegetation. Microscopic charcoal (as a record of fire) and historical accounts of vegetation supplemented pollen analysis in this comparison of inferred vegetation changes.

The Sites

Pineo Pond (67°56'19" W, 44°43'6" N), located on the Pineo Ridge delta (Fig. 1), was the principal study site. It is a kettle lake with no inlets or outlets at present. The maximum water depth is 15 m (Maine Dept. Inland Fish. Wildl. 1982). Slopes of greater than 45 degrees directly surround most of the 3-hectare (ha) basin. To the west is a large gully with a gentler slope in which surface drainage occurs seasonally. To the east is a forested saddle that sits below the delta plain and leads from Pineo Pond to a series of other ponds that drain to a boggy area toward the north. (Erosion marks high on the slopes around Pineo Pond make it probable that at some point in the past this pond was connected through this saddle to the other ponds.) To the north, the slopes of Pineo Pond are mostly open with scattered northern red oaks and red pines and an ericaceous ground cover. The slopes to the south are more closed and are dominated by white pine and red spruce. Above, on the delta surface, more than 400 ha of commercial blueberry lands surround the pond on three sides stretching to the northwest and southeast.

The vegetation, soils, and topography around Pineo Pond contrast sharply with those around the secondary site, Mud Pond (68°5'20" W, 44°38'2" N). This 1-ha pond is located about 10 km southwest (Fig. 1), across the Narraguagus River on the forested land of the Tunk Lake granitic batholith (Maine Geol. Surv. 1967), where typical northern mixed hardwoods predominate. The maximum water depth is 15.25 m (Maine Dept. Inland Fish. Wildl. 1982). The pond lies in a bedrock depression just south and below the cliffs of Tunk Mountain with a small, usually seasonal stream entering the pond from the west. The nearby vegetation is an association of oaks, American beech, maples, birches, balsam fir, and spruces. Red spruce dominates the mountain above the pond, whereas nearer the pond is a greater proportion of hardwoods.

The two sites were selected according to the criteria of Jacobson and Bradshaw (1981). Both are small basins, so that airborne pollen influx should be predominantly local as opposed to regional in origin (Janssen 1966). The porous soil and lack of current surface inlets at Pineo Pond suggest that most pollen would be of this local airborne origin, though during periods of vegetative openness a relatively greater percentage of regional pollen may be expected. Sediments from Mud Pond may contain a more significant component of regional waterborne pollen and erosional material carried into the pond by the inlet stream and expected slope-wash from the pond's extensive and steep watershed on Tunk Mountain.

Methods

The primary data for this study were gathered from analyses of fossil pollen and microscopic charcoal contained in the sediments of Pineo and Mud ponds. Sediment analysis also included lead- (Pb-) 210 dating, dry weight, and loss-on-ignition (LOI). Additional information was derived from a review of survey data and other historical records pertaining to the structure and change in vegetation during historical times.

Sediment Sampling and Analysis

The top 55 centimeters (cm) of sediment of Pineo Pond were collected with a freeze sampler at 15 m water depth on 2 February 1985. Approximately 1 cm from the inside and 0.5 cm from the outside of an 8-cm-wide section of frozen crust were discarded to remove sediments that might have been contaminated or distorted. The remaining frozen sediment was sectioned with a saw into 0.5-cm portions. Although not properly a "core," this frozen sample is referred to as such in this paper.

At Mud Pond a 10-cm-diameter Davis and Doyle (1969) piston corer was used to obtain a 52-cm-long surface core. The core was collected in 15.2 m of water on 2 May 1985. It was sectioned in the field into 0.5-cm increments for the top 29.5 cm and into 1-cm increments from 30.0 to 48.0 cm. The bottom 4 cm were discarded because of possible contamination.

Samples were processed for pollen analysis according to standard laboratory techniques with potassium hydroxide (KOH), hydrofluoric acid (HF), and acetolysis (Faegri and Iversen 1975). From 5 to 10 tablets of club moss (*Lycopodium*) spores were added prior to processing to allow calculation of pollen concentration (Benninghoff 1962). Samples were sieved through a 150-micrometer (µm) Gruce crucible and an 80-µm sieve following the KOH treatment. At least 300 pollen grains were counted at each level. Identifications followed Richard 1970, McAndrew et al. 1973, Faegri and Iversen 1975, and Moore and Webb 1978, with reference to the University of Maine pollen collection. Routine counting was at 500 power (x) with a Leitz Ortholux microscope; more difficult grains were identified under oil immersion at 1250 x.

Pine grains were identified as white pine or red/jack pine types, although jack pine does not currently grow near either site. Undifferentiated pine grains were distributed proportionally among one of these two types. Fragments of saccate grains of conifers were identified in units of one-quarter whole grains, which were distributed proportionally among all conifer types with saccate pollen grains.

Myricaceae were divided into sweet fern and sweet gale (*Myrica*) types. Sundberg (1985) found sweet fern pollen to be distinct from sweet gale pollen. Consistent differences in pollen structure were found between the two species, including differences in endopore and ectopore diameter and overall shape.

Fire history was inferred by analyzing microscopic charcoal using the point-count intercept method (Clark 1982). This allowed calculation of total charred particle accumulation based on a statistical measure of the percentage of area of charcoal on the slide, the area of the slide counted, the pollen sum, and the pollen concentration. A particle was identified as charcoal only if it was angular and dark tan or black (Patterson et al. 1987). In many cases, the structure of the wood could still be seen.

Historical Review

Information on pre-European and nineteenth-century vegetation patterns was obtained from early survey records and other historical documents. Survey field books from this period gave general descriptions of the land, its quality, and its potential use. Records of agricultural activity and population for the area were found in various histories and state reports. Unlike for some areas in Maine, these data were not sufficient for statistical interpretation of the vegetation at the time of European settlement. The data did, however, reveal the variation of vegetation on the landscape and certain essential details of historical vegetation dynamics.

Chronology

An experimental design based on paired small basins requires an accurate chronology that establishes precise time-stratigraphic correlations between the pollen diagrams. In this study the chronology to approximately 1850 was provided from Pb-210 dating done by the University of Maine Environmental Sciences Lab (ESL). This was evaluated by comparison with the two pollen diagrams, which show changes correlated with vegetational dynamics known from the historical record. Charcoal and pollen stratigraphy were used to match the locations between the two cores. The resulting cross-correlations were then compared to regional pollen signals from other sites. In addition, radiocarbon dates were used to infer the date at the bottom of each core.

Difference Diagram

After the establishment of a sediment chronology, a difference diagram was generated by subtracting the unmodified pollen percentage curves of Mud Pond from those of Pineo Pond; alders (*Alnus*) were excluded from the pollen sums of both lakes because of the large influx of this taxon following fires at Mud Pond. The inferred ages of all samples for both ponds were used in this calculation, with linear interpolation being used to generate data when a date for one pond fell in between two sample dates for the other. This method differed from that of Jacobson (1979) who used both log-ratio values and influx values of pollen curves averaged over subintervals. Resulting positive sums (to the left of the vertical zero axis; see Fig. 6) revealed pollen types with greater percentages at Pineo Pond; negative sums (to the right of the axis) revealed pollen types with greater percentages at Mud Pond. Values of zero, or that fluctuated about the axis, indicated either no significant percentage difference or regional noise (variation unrelated to local vegetation differences between the two watersheds).

Results

Settlement History

Economic activity in eastern coastal Maine following early settlement in the late seventeenth and early eighteenth centuries was based on lumber, with little agricultural activity (Drisko 1904, Allis 1954). Because of the demand for lumber and ship masts, white pine was the first species logged (Smith 1972). The amount of wood cut increased steadily from the mid-1700s through the mid-1800s, with a two-thirds decrease in harvest between 1850 and 1882. White pine, once easily

Vegetation and Fire on Maine Grassland Barrens

available, became hard to find by about 1825; the emphasis shifted to spruce, and by 1860 more spruce was cut than pine (Smith 1972). By 1885 there was little marketable timber left to cut near the Narraguagus River (Smith 1972). The logging of eastern hemlock began in the 1830s (Black House Papers undated) and increased as the tanning industry gained importance. Harvesting peaked during the 1870s and 1880s (Allis 1954) but came to an abrupt halt about 1890 following the development of artificial extracts (Smith 1972).

Regional agriculture began by about 1820 and increased after 1840. The most intense period of agricultural activity was in the 1870s (Day 1963), when wheat production in Washington County increased more than fourfold (Drisko 1904). This activity decreased over the next 20 years but increased again after the turn of the century. Tractors, introduced after World War I, reduced the number of farms but resulted in a continual increase in the size of farms and the percentage of land in cultivated crops, as opposed to hay (Day 1963). World War II brought another major decrease in agriculture.

Survey Records

According to a variety of land surveys and other observations, the Pineo Ridge delta was somewhat open at the time of European settlement and included areas of extensive pine forests or scattered trees. Evidence of recent fires was also provided. Surveyors Joseph Pierpont and William Albee surveyed Washington and Hancock counties extensively in the fall of 1792. The survey had no witness-tree data, only descriptions of vegetation. Comparisons with modern topographical maps show that the survey was hampered somewhat by inaccurate course and distance, and perhaps by a desire to convey the best impression of the land. In general, though, it did seem to provide an accurate description of the landscape at the time of early settlement. Pierpont and Albee (1792) reported that on the east side of the Narraguagus River (toward Pineo Pond) the growth was killed by "the fire," whereas on the west side (toward Mud Pond) it was well timbered. Notes made when the two men crossed the delta surface a bit north of Pineo Pond described "a large sandy plain [with] some scattering pines and birch bushes." They wrote in their survey book that "it has the appearance of once being covered with large pines destroyed by fire" (Pierpont and Albee 1792).

Later Pierpont and Albee described a piece of land now known as the Epping Plain, 5 km southeast of Pineo Pond, as a "clear level sandy plain which is 4 miles wide. North of this plain the growth is pitch pine [*Pinus rigida*] and south of it the land is low and level [with] spruce, white pine, white [paper] and yellow birch [*Betula lutea*] and some hemlock and small cedars [*Thuja*]. The plain is the most remarkable in the County, it is capable of parading and maneuvering 40,000 Men—Excellent for a Horserace and no doubt it will be a piece of Ground of great consequence in some future time" (Pierpont and Albee 1792). The "pitch pine" they mention is most likely red pine; pitch pine does not currently grow on or near the delta, whereas red pine is common there.

Alexander Baring, a British surveyor who toured Washington and Hancock counties to evaluate the Bingham lands in eastern Maine, noted that "seashore lands produce spruce and pine [while] in the second townships the hardwoods commence and are principly oak [*Quercus*], hemlock, maple, larch [*Larix*], ash [*Fraxinus*], and elm [*Ulmus*]" (A. Baring *in* Fischer 1954). Like Pierpont and Albee, Baring also visited the part of the delta known as the Epping Plain. He described it as "a plain two or three miles in diameter very poor and barren. The soil is perfectly barren and covered with a short kind of heath and no wood. It has the appearance of having been burned, but the soil is so hard that it can never have been good....The nature of the whole is singular and different from anything I ever saw. When you get over the plain, which is a hill with a table top, you descend again on good land" (A. Baring *in* Fischer 1954). Surveyors in southern New England frequently considered "barren" land useless, when in fact it was treeless as a result of long-term burning by Native Americans (Pyne 1982).

Many fires burned throughout much of Maine and neighboring New Brunswick, Canada, in 1825. The Great Miramachi Fire of that year did not appear to have burned in the study area (Fox 1985), although Cary (1884) reported that a fire burned from Wesley to Jonesboro in 1827. The survey by Adison Dodge in 1832 clarified which part of the study area was affected (Dodge 1832). He reported that the land west of Little Long Pond and Tunk Lake (just east and south, respectively, of Mud Pond) was unaffected by the fire, whereas the land east of these ponds was burned. When Dodge attempted to recover the line run in 1824 between Cherryfield and Steuben, the first half of the line east to the Narraguagus River was burned and only a few trees remained in swampy areas (Dodge 1824). Additionally, the land north of the lots that Dodge surveyed in the southern part of Cherryfield was mostly "burnt" and "worthless," so he did not divide it into lots.

In 1857 the U.S. Coast and Geodetic Survey surveyed and built for triangulation purposes a 9.6-km baseline across the Epping Plain. It described the land as "an extensive pine barren" consisting of "rolling sand ridges covered with grass and underwood with a few clumps of trees." Patches of forest existed in the swampier places (Bache 1857). Surveyor C. O. Boutelle had undertaken a reconnaissance of the region in 1855 to find a location for the line. He described the land as generally level and open and mentioned scattered pines. While taking a southeast traverse across the delta surface near Pineo Pond, he said that the land was "level for 1/2 mile or more on either side, covered with low shrubs and a few trees" (Boutelle 1855).

Lowbush Blueberry Production

The open nature of this area in these early times may have been a result of soils unsuitable for tree growth, recent natural fires, early logging, or perhaps anthropogenic fires started by Native Americans to encourage berry production and improve hunting and ease of travel, or for other purposes. A recent, albeit limited, archeological survey turned up evidence of Native American occupation around the numerous springs that emerge from the slopes of the delta edge (Petersen and Heckenberger 1987), and blueberries are known to have been a favored food of Native Americans in eastern Canada after European contact (Arnason et al. 1981). Naturalist John Josselyn (*in* Day 1953) said that the Native Americans in southern Maine dried blueberries in the sun and sold them to the English "by the bushel." Although the local lore is that Native Americans did burn the barrens for blueberries, I found no solid documentation of it.

No early observers mentioned using the delta as a source of blueberries. Commercial blueberry packing began by 1866 (Day 1963), but harvesting for home use began before this. As late as 1871, however, the land was still considered "worthless" for taxation purposes (W. Underwood *in* Washington Co. Supreme Judicial Court 1873). The value of blueberry lands was recognized in the town of Columbia in 1882, when the townspeople voted to "tax blueberry lands according to its value and income according to other property" (Green and Drisko 1976).

In 1867 William Freeman, Jr., an attorney who had gained control of most of the potential blueberry land in the area, printed and distributed a notice that was the first attempt to restrict public harvesting of blueberries. He admitted that for "many years past" the public had harvested berries there but complained that he had lost thirty thousand dollars worth of berries in 1866 and that the fires set from year to year had destroyed much timber (W. Freeman *in* Washington Co. Supreme Judicial Court 1873).

The Freeman notice preceded a court case that eventually ended using the barrens as public commons (Pulsifer 1877). In depositions before the court in 1873, Aaron Allen, age 61, stated that he had harvested berries for 40 years, and Abram Tucker, age 40, said he had harvested berries for 24 years (Washington Co. Supreme Judicial Court 1873). This puts the first historically confirmed blueberry harvesting on the barrens in 1833. It is probable that the fire in the previous decade greatly expanded the growth of blueberries and encouraged more extensive harvesting. If not before, by 1871 extensive commercial harvesting definitely took place on the delta surface by Pineo Pond. Many depositions in this court case mentioned camping at the Porcupines (2 km northwest of Pineo Pond) and at Pikes Brook (2 km west of Pineo Pond). James Plummer, a blueberry buyer, specifically stated that his pickers were on "quite a smooth piece of barren land [called Pineo Field] beside a small pond" with no inlet or outlet. He estimated that this field was about 1.2 km southeast of the Porcupines (Washington Co. Supreme Judicial Court 1873). The pond is likely Pineo Pond.

Stratigraphic Record

Chronology

The early chronology for Mud Pond was based on Pb-210 dating performed by the University of Maine ESL. At Pb-210 background levels, the sediment at 11.5 cm was estimated to be 160 years old. A radiocarbon date (PITT-0804) obtained from sediments of a 20-cm-diameter long-core provided a date of 1,700 ± 115 years before present (YBP) for the bottom of the core (48.0 cm) used in this study. Dates were linearly extrapolated between the settlement horizon and the radiocarbon date.

Although the ESL also analyzed Pb-210 for Pineo Pond, the calculated dates were rejected as being inconsistent with known historical vegetation changes. Instead, the chronology for the Pineo Pond core was based on correlation of the distinct settlement horizon at a depth of 18.5 cm with that found in the Mud Pond diagram. A radiocarbon date (A-7050) of 1,800 ± 65 YBP was obtained at a depth of 52 cm with sediment from a 10-cm-diameter long-core and was correlated through pollen and sediments at 54 cm. The following zonations were based on stratigraphic comparison of the two cores in conjunction with the established dates.

Zonation

A. Early Period (Eastern Hemlock Decline): 1,800–1,350 YBP

This zone was marked by large charcoal values relative to background levels in each core (Figs. 2 and 3) and was estimated at 1,700 YBP. This fire was catastrophic to the regional vegetation and apparently was responsible for major declines in both eastern hemlock and American beech percentages (Figs. 4 and 5). The pollen abundances of major forest taxa such as pines and birches were reduced by the fire and replaced in the record by shrubs and herbs. The tree species showed recovery by the end of the period. Spruces started an increase that persists to the present day. Based on the radiocarbon date, the bottom of the Pineo Pond core was estimated to have been deposited about 1,800 YBP, making it slightly older than the bottom of the Mud Pond core, which provided a date of about 1,700 YBP. It was apparent from the higher eastern hemlock levels and greater time below the fire event that the bottom of the Pineo Pond core was older.

This fire resulted in an opening of the Pineo Ridge vegetation, with pollen spectra similar to that occurring at the time of European settlement. Most major open-indicators, including the Gramineae, Ericaceae, sweet fern, and bracken, increased (Fig. 2). Both white pine (shown in dark) and jack/red pine (shown in outline) pollen types also increased, suggesting a strong secondary successional response. At Mud Pond, alder percentages increased greatly following the fire (Fig. 3). Some of this increase was likely due to regrowth of the shrub border around the pond.

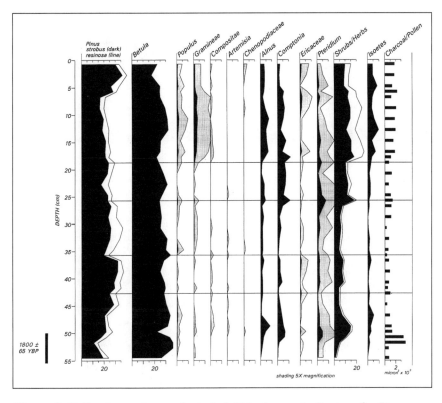

Figure 2. *Pollen percentages of selected disturbance indicators for Pineo Pond.*

B. White Pine Zone: 1,350–1,000 YBP

This period was defined by initial high percentages of white pine and birch in both ponds. White pine slowly increased in the Pineo

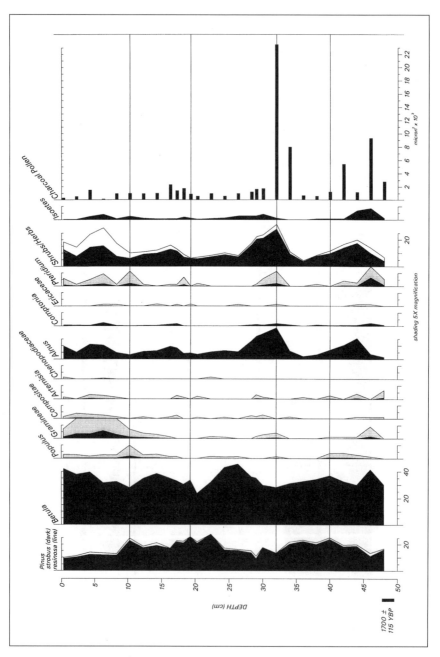

Figure 3. Pollen percentages of selected disturbance indicators for Mud Pond.

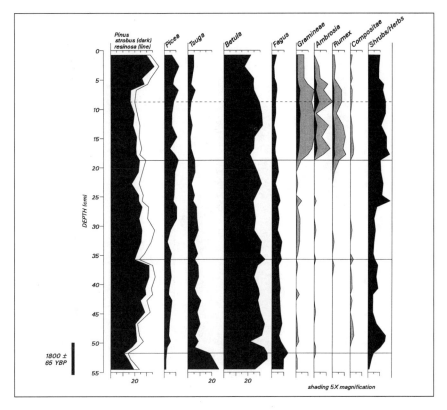

Figure 4. Pollen percentages of selected regional indicators for Pineo Pond.

Pond core, whereas in the Mud Pond core it appeared to be replaced at first by spruce, eastern hemlock, and American beech. This resulted in an increasing difference in the percentage of white pine between the two cores. Birch, however, showed a gradual decrease in both ponds until the end of this time period.

A catastrophic fire apparent in the Mud Pond diagram at the end of the period resulted in the reduction of all percentages of tree pollen and a major increase particularly in alder pollen. This fire may have been the same as that shown in the Pineo Pond diagram at about this time (illustrated by increased sweet fern and bracken rather than increased charcoal levels).

C. Pineo Red Pine Zone: 1,000–500 YBP

This period was defined by high percentages of jack/red pine in the Pineo Pond diagram. Each core showed a coincident decline of about 40 percent in white pine percentages, reaching the lowest value prior to settlement. Spruce rose to a peak in both ponds about 500 to 700 years ago (Figs. 4 and 5); the higher percentages in the sediments of Pineo Pond compared to Mud Pond probably indicate a more open

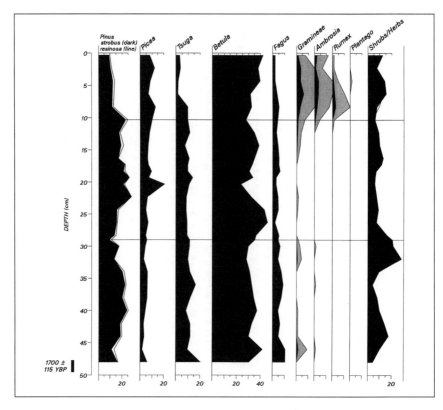

Figure 5. Pollen percentages of selected regional indicators for Mud Pond.

vegetation. A fire occurred at the top of the zone 1,000 YPB, shown by a slight charcoal increase in the Mud Pond core, a decline in forest species, and an increase in successional species.

D. Early Colonization Period: 500-160 YBP

The fire at the zone boundary estimated at 500 YBP resulted in changes in both diagrams that were similar to, though less intense than, changes 1,000 and 1,700 YBP. Pollen percentages of pine, spruce, and eastern hemlock decreased, whereas pollen percentages of birches, Gramineae, poplars (*Populus*), bracken, and other distinctive early-successional species increased. The decline of white pine as the dominant forest species probably occurred first in response to this fire and then second because of regional cutting of mature trees by early settlers. Percentages of this taxon then increased at the top of the zone in the Pineo Pond core. Birch percentages showed a successional increase and then a decrease as a mature forest developed.

E. Agricultural Settlement Period: 160 YBP-Present Day

The settlement period has been marked by increases in agricultural indicators such as Gramineae, ragweeds (*Ambrosia*), and dock

(*Rumex*) and other herbs. The Pineo Pond diagram showed two important periods of agricultural activity, centered about 140 years ago (1855) and about 65 years ago (1930). This is consistent with the agricultural history of the Narraguagus River valley. Mud Pond showed a single peak in activity about 85 years ago (1910). This apparent contradiction can be reconciled in part by the differences in resolution of the two diagrams (being much higher at Pineo Pond), but it likely reflects differences in agricultural activity near the two ponds.

The pollen percentages of the major timber species fluctuated consistently within each core. Both showed declines in white pine, spruce, and eastern hemlock, with low percentages reached by the end of the nineteenth century. The percentage of birch pollen increased in response to forest clearance throughout the beginning of the period.

Pineo Pond showed high levels of charcoal with no distinct peaks. Sweet fern, bracken, and ericaceous pollen types were also abundant throughout most of the period. This is consistent with the extensive historical burning and openness associated with blueberry management.

Difference Diagram

A comparison of diagrams from the two cores revealed marked differences in vegetation history as shown in the two stratigraphies (Fig. 6). Pineo Pond showed higher levels of red pine, sweet fern, and bracken and a more consistent presence of Gramineae. This suggests that these taxa were of greater importance not only in the historical period but throughout the time period represented by the cores. Mud Pond showed more eastern hemlock and spruce (prior to cutting in the nineteenth century), birch, and maple, with a few notable exceptions. During the early settlement period and after the fire about 1,700 YBP, the growth of birch seemed to overwhelm the spruce inputs at Mud Pond. Because of its dominance at Mud Pond following the fire, alder was excluded from the pollen sum used to calculate the diagram.

The difference diagram also suggests vegetation dynamics not readily apparent in the individual diagrams. The white pine period (1,350–1,000 YBP), for example, appeared to be largely dominated by white pine at Pineo Pond. The difference diagram suggests that the situation was more complex. The oscillation of sweet fern and bracken against rising white pine values within this period suggests that a significant portion of the pine pollen may have been a result of long-distance transport or local overproduction. Ericaceous pollen also was consistently more abundant in the Pineo Pond diagram (Fig. 2), indicating that the delta might have had a more open component than the high pine percentages would suggest. This is supported by other analyses indicating a decrease in sediment concentrations of pine pollen in this period, with sweet fern and bracken levels both increasing.

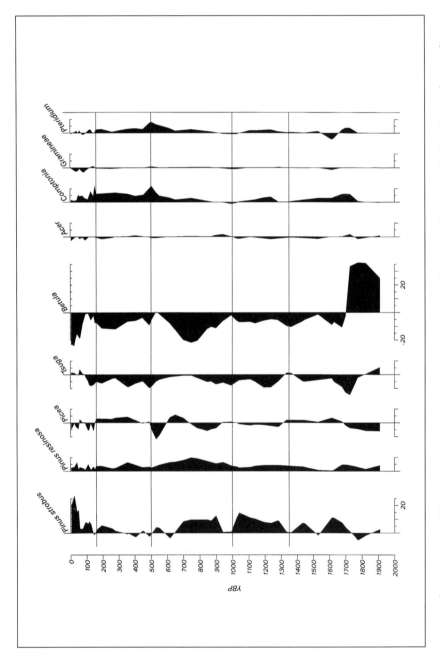

Figure 6. Percentage difference between selected pollen types for Mud and Pineo ponds. Positive values indicate more of that type at Pineo Pond.

Discussion

Regional Vegetation Changes

The overall trends of eastern hemlock, American beech, and spruce pollen in the Pineo and Mud pond diagrams were generally consistent with the behavior of these taxa in Maine and neighboring Canada over the past 1,000 to 2,000 years (Mott 1975, Gajewski 1983, Anderson et al. 1986, Jacobson et al. 1987). These changes have been interpreted as evidence of a cooling climate, with the ecotone line between the boreal forest and the northern hardwoods moving southward (Gajewski 1983, Anderson et al. 1986). Eastern hemlock and American beech pollen percentages decreased in association with a charcoal peak about 1,700 years ago in both the Pineo and Mud pond diagrams. This date was within the period of a similar decline in hemlock percentages in dated diagrams from Upper South Branch Pond (Piscataquis Co., ME; Anderson 1979, Anderson et al. 1986), Gould Pond (Hancock Co., ME; Jacobson et al. 1987), and Conroy Lake (Aroostook Co., ME; Gajewski 1983) and from lakes in southwestern Nova Scotia (Mott 1975). Spruce percentages increased stepwise at Mud and Pineo ponds throughout the period studied, with a particular increase during the Little Ice Age period starting about 500 years ago (Lamb 1969).

The curves of these species appear similar for the two ponds, although those from Mud Pond seem to have been influenced by local fire events. Neither eastern hemlock nor American beech grows today on the deltas near Pineo Pond, and spruce occurs only sparsely on the north-facing and more sheltered slopes of the pond. It is likely that the present soils of the delta and the fire regime shown through the period studied have kept these species from becoming very common on the delta. These species curves at Pineo Pond appear to represent the integrated regional behavior of the species, whereas at Mud Pond the curves are more representative of local abundances.

Fire Regime

The role of fire appears to have differed at the two sites. Until the settlement period, charcoal/pollen ratios at Pineo Pond appeared low when compared to the major peaks at Mud Pond. Even during the settlement period, when fires were set for management purposes every two or three years, charcoal levels were not particularly high in the Pineo Pond data. The peaks that occurred tended to be associated with peaks of bracken and sweet fern but often were not much above background level.

Background charcoal/pollen ratios (charcoal area/pollen grain) were high at Pineo Pond, however, with levels generally ranging between

400 and 700 square μm per pollen grain (Fig. 2). This was similar to peak values found for The Bowl and Sargent Mountain Pond on nearby Mount Desert Island in Hancock County (Patterson et al. 1983, 1987). These persistent high levels of charcoal, combined with the constant presence of fire-adapted species, indicate a regime on the Pineo Ridge delta of relatively frequent but generally noncatastrophic fires over most of the period covered by this study. Occasionally a more intense fire may have opened what had become a closed, mostly pine-dominated forest. Prior to the fire 1,700 years ago, the Pineo Ridge delta appears to have had extensive hemlock growth that was destroyed in a major regional fire. Five to six fires of medium to high intensity appear to be recorded in the diagram in the 1,500 years prior to agricultural activity, indicating a time interval of about 250 to 300 years between major fires. It seems probable, however, that many more fires of a smaller or less intense nature occurred on the delta.

Quite a different story was revealed at Mud Pond. The background levels of charcoal were lower than at Pineo Pond, particularly after settlement time. Charcoal peaks were much higher, however, and were distinct from general background levels. Pollen percentages following fire at Mud Pond were dominated initially by alder instead of by sweet fern, bracken, or pine, as was recorded in the Pineo Pond core. Alder in the Mud Pond core was replaced by successional increases in birch and then spruce and hemlock. This indicates the different forest type and moisture regime at Mud Pond. The difference in charcoal peaks can probably be explained by the more catastrophic nature of fires in the Mud Pond watershed, greater charcoal production resulting from a larger initial biomass, and differing bathymetries of the basins. The data show that fire in the Mud Pond watershed resulted in greatly increased erosion, which would be expected to have added a large component of waterborne charcoal to the pond. In contrast, fire at Pineo Pond resulted in much less erosion and therefore a greater proportion of airborne charcoal. In fact, the ratio of charcoal to inorganic matter was higher overall at Pineo Pond.

Three principal fires are evident in the Mud Pond diagram in the 1,500 years prior to settlement and two in the 600 years prior to early lumber activity. This indicates a fire-rotation period of about 300 years. Smaller fires of any significance appear to have been unlikely between these major fires. This is generally longer than the fire rotation indicated by the Pineo Pond data. The fires that did occur caused catastrophic changes in the vegetation.

Historical and Pollen Records

The upper part of the pollen record from Pineo Pond was generally consistent with known historical events. The pine growth and extensive clearings documented in survey records made on the delta prior to extensive European use 200 years ago were confirmed by the pollen record. Early clearing and burning on the Pineo Ridge delta were confirmed by the increase of sweet fern, bracken, and charcoal in the core. Increased fire frequency following settlement was shown by increased charcoal levels. This burning increased the ericaceous component of the pollen record and decreased the abundance of sweet fern and bracken, which are weed species in blueberry cultivation. Extensive openings appeared on the barrens within the past 150 years. Since the mid-1940s, "weed" shrub species have been further reduced through intensive burning and, more recently, herbicide use, as seen near the top of the core. The high pine values found in the upper core were probably a result of high local production near the pond and of long-distance transport over the open landscape of ericaceous barrens.

Clear indications are seen in both diagrams of the results of timber cutting. Regional cutting of white pine appears as early as 300 years ago, though the greatest percentage drop prior to local agriculture is about 260 years ago. Local activity is represented by the drop in white pine about 150 years ago, consistent with a date of 1820 found at The Bowl on Mount Desert Island (Patterson et al. 1983). Spruce and eastern hemlock declines are seen in the 1800s, with low percentages of these species by the later part of that century when most available timber in the area had been cut.

Fire in the Prehistoric Era

The vegetation and fire histories inferred from the Pineo Pond pollen diagram support the hypothesis of a landscape that was ecologically adapted to fire. The vegetation of the Pineo Ridge delta, whether shrubland or pine forest, would have been easy to ignite during periods of drought. In fact, in any normal year following snow melt in the spring and before leaf-out, a typical blueberry field will ignite readily (JCW pers. obs.). The flat topography of the delta provides few fire breaks, and once ignited, a fire would likely spread to the downward slopes of the delta edge.

Changes similar to those that occurred at the time of settlement appear in the Pineo Pond diagram as far back in time as this short core provides data. Fires resulted in increased open areas, establishment of ericaceous and sweet fern shrublands (as occurs in abandoned blueberry fields today), and the eventual development of a pine forest or woodland. Although the Pineo Ridge delta has become more open during historical times, these shrubland habitats seem to have been partially persistent. Sandy soils of the delta were frequently xeric. The ignition-prone vegetation, high background charcoal levels, and cyclical evidence of fire-initiated openings suggest a primary role for fire in maintaining the vegetation.

The vegetation and topography near Mud Pond, however, appear to have inhibited frequent ignition and spread of fires. Numerous lakes and a steep topography provide many fire breaks. For example, the fire noted by Pierpont and Albee (1792) burned the east (Pineo Pond) side of the Narraguagus River but did not cross it. The fire noted by Dodge (1832) reached only as far as Tunk Lake, 3 km east of Mud Pond. Still, two major fires did occur in the Mud Pond watershed in the 600 years prior to settlement, giving an estimated fire-return time of about 300 years.

Vegetation and Fire

Ericaceous species are adapted to fire. They have a well-developed root system, with adventitious rhysomatic roots that allow lateral vegetative growth to develop large clones. Older plants have a taproot up to 1 m deep that aids in fire and drought resistance. When grassland barrens are established by burning, they often take a long time to grow back to forest. These areas have a natural tendency to burn again, with succession often stopping at an open pine woodland (Dansereau 1957). Sweet fern, a prolific pollen producer and a major indicator of open areas in this study, is a common associate in blueberry fields and is also aided by fire (Hall et al. 1973). Dense thickets often form in acid, sandy soil (Fessenden et al. 1973). The presence of these taxa in the pollen record is therefore consistent with a high fire frequency.

Under certain circumstances, persistent disturbance can locally eliminate tree species and promote the establishment of stable shrubland habitats that can resist reinvasion of trees (Niering and Goodwin 1974). These plant associations may vary in composition, but they often include ericaceous and other shrubs such as sweet fern, blueberries, and sheep laurel. Sheep laurel in particular can cause changes in soil fertility that are unfavorable to tree growth (Damman 1971).

Sweet fern seeds have been shown to survive at least 40 years in soil and to sprout only when the overstory is cleared (Tredici 1977). Clearing by fire produces denser sweet fern stands than clearing without fire (Lutz and Cline 1956). This nitrogen-fixing shrub may play a similar role in this ecosystem to what pin cherry (*Prunus pensylvanica*) does in northern hardwood forests (Marks 1974). Fire clears the forest, releasing latent sweet fern seeds. The emerging plants trap nutrients and become nodulated with *Frankia*, a nitrogen-fixing bacteria that has been shown to fix atmospheric nitrogen (Ziegler and Hüser 1963). These communities can persist in areas of frequent fires, such as in the jack pine forests of Canada (Ziegler and Hüser 1963).

Causes of Fire

It is difficult to determine the causes of ignition for fires recorded in the sedimentary record. The fires of the late eighteenth and early nineteenth centuries could be natural (e.g., lightning) or anthropogenic (e.g., slash burning by early settlers; Allis 1954). The majority of fires in historical times were caused by human activity. Only 15 percent of the fires in the early part of the twentieth century in northern Maine were caused by lightning (Fobes 1944). During this century in neighboring New Brunswick, only 7 percent of fires could be attributed to lightning (Wein and Moore 1977), and in Nova Scotia only 1 percent of fires were lightning caused (Wein and Moore 1979). Patterson et al. (1983) found that lightning caused even fewer fires near the coast than inland.

The low frequency of natural lightning fires in Maine, particularly along the coast, leads to consideration of the role of Native Americans in fire ignition. The persistent nature of fire on the Pineo Ridge delta during most of the last 900 years suggests a fire frequency higher than that attributable to natural ignition alone.

Conclusion

The pollen data for Pineo and Mud ponds show similar records of regional vegetation change that are consistent with historical records and previous paleoecological studies. The data clearly demonstrate that the two ponds differed in the importance of the predominant species and in the inferred fire regime. The Pineo Ridge delta was opened by a major fire about 825 years ago. Since that time the pollen record indicates higher percentages of species such as red pine, sweet fern, and bracken, all taxa that are adapted to dry conditions and frequent fires. The high percentages of these fire-adapted species of trees and shrubs combined with high charcoal levels indicate a frequent return of low-intensity or small localized fires on the delta surface. Mud Pond has had higher percentages of more mesophytic, fire-intolerant species such as eastern hemlock, American beech, and spruce. The fires at Mud Pond had a catastrophic effect on the vegetation, which was generally not true of the fires at Pineo Pond.

The difference in soils at the two sites does not explain the differences in vegetation. The topography and drought-prone soils of the Pineo Ridge delta favored some xerophytic species and increased the likelihood of fire at that location; however, fire was probably the major factor affecting the vegetation.

Interesting questions remain regarding the areal extent of fires through time on the Pineo Ridge delta. Did the entire delta have similar trends of vegetation history, or were spatial and chronological heterogeneity present? Survey records indicate that at the time of settlement the entire delta was a mosaic of differing vegetation types. The paleoecological evidence shows consistent (though varying) evidence of openness on the delta. This evidence throughout the time period studied is consistent with either (1) a homogeneous landscape of more open pine-woodland made occasionally more open by fire, or (2) the shifting-mosaic hypothesis (Heinselman 1973), with a resultant landscape of fire-created openings surrounded by more closed pine forest.

It seems likely that certain fires (i.e., those 1,700 YBP) burned most if not all of the delta, whereas smaller, more localized fires were the rule during other periods. The first scenario may be more consistent with the hypothesis that sandy soils and topography along with natural ignition were sufficient to promote the fire regime of the Pineo Ridge delta. The second lends support to the hypothetical role of humans, who would have been likely to start small localized fires under somewhat controlled conditions, in promoting fire ignition. This point would be clarified by studies involving a suite of sites across the complex of deltas found in the area, perhaps combined with a detailed fire-prediction model based on topography and vegetation. This might help test some of the inferences suggested by the results of this study.

This study has provided clear evidence that the open nature of the Pineo Ridge delta was not due solely to historical management practices. Even during periods of high overall tree pollen percentages, the evidence suggests that at least parts of the delta near Pineo Pond were open for long periods during the period studied. At a minimum, shrubland openings have recurred near the pond at intervals of about 150 years for at least the last 900 years.

Literature Cited

Allis, F. S. 1954. William Bingham's Maine lands:1790-1820. Colonial Soc. Massachusetts, Boston.

Anderson, R. S. 1979. A Holocene record of vegetation and fire at Upper South Branch Pond in northern Maine. Master's thesis, Univ. of Maine, Orono.

Anderson, R. S., R. B. Davis, N. G. Miller, and R. Stuckenrath. 1986. History of late- and post-glacial vegetation and disturbance around Upper South Branch Pond, northern Maine. Can. J. Bot. 64: 1977-1986.

Arnason, T., R. J. Hebda, and T. Jones. 1981. Use of plants for food and medicine by native peoples of eastern Canada. Can. J. Bot. 59: 2189-2325.

Bache, A. D. 1857. A base of verification upon the Epping Plains. Coast and Geodetic Survey, RG23:841GB, acc. #25982, Natl. Archives, Washington, D.C.

Benninghoff, W. S. 1962. Calculation of pollen and spore density in sediments by addition of exotic pollen in known quantities. Pollen et Spores 4: 332-333.

Black House Papers. Undated. Maine State Archives, Augusta.

Borns, H. W., Jr. 1979. Emerged glaciomarine deltas in Maine. Maine Critical Areas Prog., State Planning Office, Augusta.

Boutelle, C. O. 1855. A reconaissance of the Epping Plain. Coast and Geodetic Survey, RG23:841GA, acc. #25456, Natl. Archives, Washington, D.C.

Cary, A. 1884. Early forest fires in Maine. Pp. 37-59 *in* Annual report of the forest commissioner of Maine. State of Maine, Dept. Forestry, Augusta.

Clark, R. L. 1982. Point count estimation of charcoal in pollen preparations and thin sections of sediments. Pollen et Spores 24: 523-535.

Damman, A. W. H. 1971. Effect of vegetation changes on the fertility of a Newfoundland forest site. Ecol. Monogr. 41: 253-270.

Dansereau, P. 1957. Biogeography: an ecological perspective. Ronald Press Co., New York.

Davis, R. B. 1966. Spruce-fir forests of the coast of Maine. Ecol. Monogr. 36: 76-94.

Davis, R. B., and R. W. Doyle. 1969. A piston corer for upper sediment in lakes. Limnol. and Oceanog. 14: 643-648.

Davis, R. B., G. L. Jacobson, Jr., L. S. Widoff, and A. Zlotsky. 1983. Evaluation of Maine peatlands for their unique and exemplary qualities. Maine Dept. Conservation, Augusta.

Day, C. A. 1963. Farming in Maine: 1860-1940. Univ. of Maine Press, Orono.

Day, G. M. 1953. The Indian as an ecological factor in the northeastern forest. Ecology 34: 329-346.

Dodge, A. 1824. Dodge field notebook #14, roll 1. Maine State Archives, Augusta.

Dodge, A. 1832. Dodge field notebook #11, roll 1. Maine State Archives, Augusta.

Drisko, G. W. 1904. Narrative of the town of Machias. Press of the Republican, Machias, ME.

Faegri, K., and J. Iversen. 1975. Textbook of pollen analysis. 3d. ed. Hafner, New York.

Fahey, J. J., and W. A. Reiners. 1981. Fire in the forests of Maine and New Hampshire, USA. Bull. Torrey Bot. Club 108: 362-373.

Fessenden, R. J., R. Knowles, and R. Brouzes. 1973. Acetylene-ethylene assay studies on excised root nodules of *Myrica asplenifolia* L. Soil Sci. Soc. Am. Proc. 37: 893-898.

Fischer, R. A. 1954. Calendar of the letters of Alexander Baring, 1795-1801. Manuscript Div., Library of Congress, Washington, D.C.

Fobes, C. B. 1944. Lightning fires in the forests of northern Maine. J. Forestry 42: 291-293.

Foster, D. R. 1983. The history and pattern of fire in the boreal forest of southeastern Labrador, Canada. Can. J. Bot. 61: 2459-2471.

Fox, C. 1985. The great fire in the woods. Master's thesis, Univ. of Maine, Orono.

Gajewski, K. 1983. On the interpretation of climatic change from the fossil record: climatic change in central and eastern United States over the past 2000 years estimated from pollen data. Ph.D. diss., Univ. of Wisconsin, Madison.

Green, D. G. 1981. Time series and postglacial forest ecology. Quat. Research 15: 265-277.

Green, D. G. 1982. Fire and stability in the postglacial forest of southeastern Nova Scotia. J. Biogeog. 9: 29-40.

Green, N. H., and C. H. Drisko. 1976. A history of Columbia and Columbia Falls. Narraguagus Printing Co., Cherryfield, ME.

Grimm, E. C. 1984. Fire and other factors controlling the vegetation of the big woods region of Minnesota. Ecol. Monogr. 54: 291-311.

Hall, I. V., L. P. Jackson, and C. F. Everett. 1973. The biology of Canadian weeds: *Kalmia angustifolia* L. Can. J. Plant Sci. 53: 865-873.

Heinselman, M. L. 1973. Fire in the virgin forests of the boundary waters canoe area, Minnesota. Quat. Research 3: 329-382.

Heinselman, M. L. 1981. Fire intensity and frequency as factors in the distribution and structure of northern ecosystems. Pp. 7-57 *in* Fire regimes and ecosystem properties (H. A. Mooney, J. M. Bonnicksen, N. L. Christensen, J. E. Lotan, and W. A. Reiners, eds.). Gen. Tech. Rep. WO-26, U.S. Forest Serv., Washington, D.C.

Jacobson, G. L., Jr. 1979. The palaeoecology of white pine (*Pinus strobus*) in Minnesota. J. Ecol. 67: 697-726.

Jacobson, G. L., Jr., and R. H. W. Bradshaw. 1981. The selection of sites for paleovegetational studies. Quat. Research 16: 80-96.

Jacobson, G. L., Jr., T. Webb III, and E. C. Grimm. 1987. Patterns and rates of vegetation change during the deglaciation of eastern North America. Pp. 277-288 *in* The geology of North America, vol. K-3 (W. F. Ruddiman and H. E. Wright, eds.). Geol. Soc. Am., Boulder, CO.

Janssen, C. R. 1966. Recent pollen spectra from the deciduous and coniferous-deciduous forests of northeastern Minnesota: a study in pollen dispersal. Ecology 47: 804-825.

Kozlowski, T. T., and C. E. Ahlgren. 1974. Fire and ecosystems. Academic Press, New York.

Küchler, A. W. 1975. Potential natural vegetation of the conterminous United States. 2d ed. Spec. Publ. 36, Am. Geograph. Soc., New York.

Lamb, H. H. 1969. Climatic fluctuations. Pp. 173-249 *in* General climatology, vol. 2 (H. Flohn, ed.). Elsevier, Amsterdam.

Lorimer, C. G. 1977. The presettlement forest and natural disturbance cycle of northeastern Maine. Ecology 58: 139-148.

Lutz, R. J., and A. C. Cline. 1956. Results of the first thirty years of experimentation in silviculture in the Harvard forest, 1908-1938. Part II. Natural regeneration methods in white pine-hemlock stands in light sandy soils. Harvard Forest Bull. no. 27.

Maine Department of Inland Fisheries and Wildlife. 1982. Lake surveys. Maine Dept. Inland Fish. Wildl., Augusta.

Maine Geological Survey. 1967. Preliminary geologic map of Maine. Maine Geol. Surv., Augusta.

Marks, P. L. 1974. The role of pin cherry (*Prunus pensylvanica* L.) in the maintenance of stability in northern hardwood ecosystems. Ecology 44: 73-88.

McAndrew, J. H., A. A. Berti, and G. Norris. 1973. Key to the quaternary pollen and spores of the great lakes region. Life Sci. Misc. Publ., Royal Ontario Mus., Toronto.

Moore, P. D., and J. A. Webb. 1978. An illustrated guide of pollen analysis. Halsted Press, New York.

Mott, R. J. 1975. Palynological studies of lake sediment profiles from southwestern New Brunswick. Can. J. Earth Sci. 12: 273-288.

Niering, W. A., and R. H. Goodwin. 1974. Creation of relatively stable shrublands with herbicides: arresting "succession" on rights-of-way and pastureland. Ecology 55: 784-795.

Patterson, W. A., K. J. Edwards, and D. J. Maguire. 1987. Microscopic charcoal as a fossil indicator of fire. Quat. Sci. Rev. 6: 3-23.

Patterson, W. A., K. E. Saunders, and L. J. Horton. 1983. Fire regimes of the coastal Maine forests of Acadia. U.S. Dept. Interior, Natl. Park Serv., Boston.

Petersen, J. B., and M. J. Heckenberger. 1987. Archaeological phase I survey and phase II testing of the laser interferometer gravity-wave observatory (LIGO) project in Columbia Township, Washington County, Maine. Archaeology Research Center, Univ. of Maine, Farmington.

Pierpont, J., and W. Albee. 1792. A journal over 1,000000 acres of land in the counties of Hancock and Washington. Handwritten journal. Historical Soc., Philadelphia, PA.

Pulsifer, J. D. 1877. Reports of cases in law and equity determined by the supreme judicial court of Maine. Freeman vs. Underwood et al. Maine Reports 64: 229-233.

Pyne, S. J. 1982. Fire in America: a cultural history of wildland and rural fire. Princeton Univ. Press, Princeton, NJ.

Richard, P. 1970. Atlas pollinique des arbres et de quelques arbustes indigenes du Quebec. Nat. Can. 97: 1-306.

Smith, D. 1972. A history of lumbering in Maine 1861-1960. Univ. of Maine Press, Orono.

Stone, G. H. 1899. The glacial gravels of Maine and their associated deposits. U.S. Geol. Surv. Monogr. 34.

Sundberg, M. D. 1985. Pollen of the Myricaceae. Pollen et Spores 27: 15-28.

Tredici, P. del. 1977. The buried seeds of *Comptonia peregrina*, the sweet fern. Bull. Torrey Bot. Club 104: 270-275.

Washington County Supreme Judicial Court. 1873. Court records of William Freeman Jr. vs. William Underwood et al. (24-404). Docket 37, Maine State Archives, Augusta.

Wein, R. W., and D. A. MacLean, eds. 1983. The role of fire in northern circumpolar ecosystems. John Wiley, New York.

Wein, R. W., and J. M. Moore. 1977. Fire history and rotations in the New Brunswick acadian forest. Can. J. Forest Res. 7: 285-294.

Wein, R. W., and J. M. Moore. 1979. Fire history and recent rotation periods in the Nova Scotia acadian forest. Can. J. Forest Res. 9: 166-178.

White, P. S. 1979. Pattern, process, and natural disturbance in vegetation. Bot. Rev. 45: 229-299.

Zackrisson, O. 1977. Influence of forest fires on the north Swedish boreal forest. Oikos 29: 22-32.

Ziegler, H., and R. Hüser. 1963. Fixation of atmospheric nitrogen by root nodules of *Comptonia peregrina*. Nature 199: 508.

J. Chris Winne: Department of Botany, University of Maine, Orono, ME 04469 (current address: Cooperative Wildlife Research Unit, University of Montana, Missoula, MT 59801).

Paleoecology and Historical Ecology of an Extensive Bluejoint Reedgrass Grassland in Coastal Eastern Maine

Ann C. Dieffenbacher-Krall

Abstract

Paleoecological and historical insights about the origin and fire history of an ecological feature can indicate the most effective management techniques to preserve its current condition. The East Stream Grassland, consisting of about 800 hectares (ha) dominated by bluejoint reedgrass (*Calamagrostis canadensis*) on rolling hills, is the most extensive example of this vegetation type known in Maine. A pollen stratigraphy from an adjacent bog contained no indicators of open, grassland conditions before the clearance of land by European settlers. Historic writings and records suggest a mid-nineteenth-century origin of the grassland by means of fire. Maine Forest Service records, forest-type maps, and aerial photographs show that shifts in the grassland/forest ecotone have been minimal since at least 1940 in spite of a lack of extensive fires since at least 1967. Fire apparently has not been the critical factor in the maintenance of a majority of the grassland but may have resulted in the creation of new grassland areas. The grassland may be maintained by the autecology of the grass itself. These results suggest that a low-input management plan may be most appropriate for the site.

This manuscript is adapted from an earlier version that appeared in Natural Areas Journal, vol. 16, no.1, pp. 3-13. Reprinted by permission of the Natural Areas Association.

Introduction

According to local tradition, drought and extensive fires in the 1850s were responsible for the initiation of the East Stream Grassland in Cutler, ME (Cates and Dennison 1976, McMahon 1992, J. Cates and N. Beam pers. comm.). Frequent fires of human origin are believed to have maintained them (Famous and Spencer 1991). However, a post-European settlement origin of the grassland is not a foregone conclusion. For example, Winne (1988) found that the Pineo Ridge barren, about 80 kilometers (km) west of Cutler, was opened by fire about 825 years ago, well before the arrival of Europeans.

Bluejoint reedgrass grasslands are unusual in Maine and are limited to eastern coastal regions of the state (Famous and Spencer 1991). These grasslands also occur infrequently in Pennsylvania, but they develop frequently in western North America, including in the western Canadian arctic tundra, southern Alaska, southern Ontario, and northern Minnesota, following clearance of forest stands by logging or fire (Mitchell and Evans 1966, Bliss and Wein 1972, Mitchell 1974, 1975, Smith and James 1978, Haeussler and Coates 1986, Hogg and Lieffers 1991b). The Maine Coast Heritage Trust, which arranged for the gift of the East Stream Grassland to the State of Maine, considers the site to have high potential as an ecological reserve because it contains examples of ecosystems that in Maine are typical of and restricted to Maine's eastern coast, specifically coastal plateau bogs, maritime spruce (*Picea*)-fir (*Abies*) forests, and bluejoint reedgrass grasslands (McMahon 1992). The management goal for the East Stream Grassland is to maintain a multiple-use site providing open space, habitat protection, and ecological diversity as well as backcountry recreation and commercial timber, lowbush blueberry (*Vaccinium angustifolium*), and balsam fir (*Abies balsamea*) tip harvesting.

The objectives of this study of the East Stream Grassland were (1) to determine the approximate date of the initial appearance of the grassland to verify the general opinion that it is of human origin; (2) to determine the factors that have been responsible for the maintenance of the grassland since its development; and (3) to determine whether prescribed burning should be a component of the long-term management plan.

Site Description

The East Stream Grassland is located primarily in the town of Cutler, Washington Co., ME (44.72°N, 67.20°W; Fig. 1). Relative to most of Maine, Cutler experiences warm winters (average minimum temperature -10°C), cool summers (average maximum temperature 17.2°C),

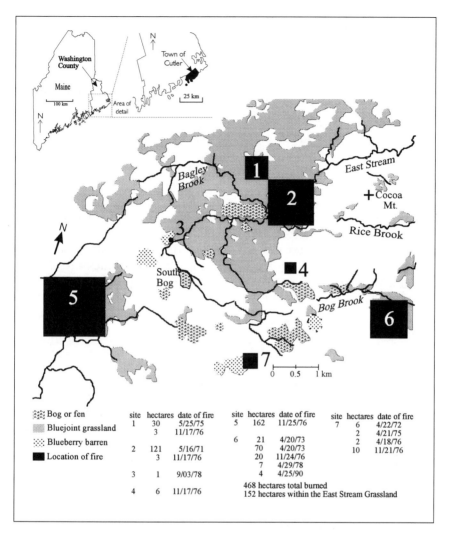

Figure 1. Fires on the East Stream Grassland since 1967 (from Maine Forest Service records). Shaded squares indicate total size and approximate location of the fires; the actual shapes of individual fires are unknown. Redrawn with permission from Dieffenbacher-Krall 1996.

and high annual precipitation (118-127 centimeters [cm] per year; McMahon 1992). The region is covered by fog for nearly twice as many hours each year as more western sections of the Maine coast (McMahon 1992).

The 4,230-ha site consists of bluejoint reedgrass grassland, streamside meadows, alder (*Alnus*) thickets, peatlands, and commercially managed lowbush blueberry barrens in the midst of spruce-fir forest (McMahon 1992). The East Stream bluejoint reedgrass grassland is the largest example of this vegetation type in Maine (about 800 ha), and it occurs on terrain that is hillier than that of any similar site (Famous and Spencer 1991). Moderately to excessively well drained stony tills, gravel, and sand predominate in the upland areas (U.S. Dept. Agric. Soil Conserv. Serv. 1983). Glaciomarine silts and clays underlie the numerous peatlands on the site. The vegetation in the East Stream Grassland continues virtually unchanged from well-drained uplands to hydric lowlands (Famous and Spencer 1991).

The grassland is characterized by low species diversity and is dominated by bluejoint reedgrass. Clumps of alder and willow (*Salix*), occasional mature white spruce (*Picea glauca*), and several even-aged stands of quaking aspen (*Populus tremuloides*; 5-10 cm diameter at breast height) dot the grassland. Young conifers are notably absent. Occasional small (< 500 square meter [m]) patches of bluejoint reedgrass also occur within the forest that surrounds the grassland.

Methods

Historical Research

Historical research sources included regional newspapers (*Bangor Daily Whig & Courier* and *Machias Union*) from 1853 to 1881, property deeds, U.S. Department of Agriculture census records of agriculture and industry (1850, 1860, 1870, and 1880), books on local history (Drisko 1904, Cates and Dennison 1976), a contemporary reminiscence (Wilder 1879), recollections of lifelong residents (N. Beam and J. Cates pers. comm.), records of the Maine Forest Service, and Cutler town records (available for 1844 to 1877 and since 1900).

Charcoal and Pollen Analysis

A core of the uppermost 1 m of peat was obtained from the deepest area (4 m) of an unnamed peatland (referred to here as South Bog) adjacent to the East Stream Grassland using a 10-cm-diameter Russian peat sampler (Jowsey 1966, Aaby and Digerfeldt 1986). The core was cut into 0.5-cm slices, and 1-milliliter (ml) samples were removed from the core for pollen analysis using standard procedures (Fœgri et al. 1989). Three conventional radiocarbon dates were obtained from bulk peat samples (see Table 1). Additional dates were inferred from a comparison of the regional pollen record from South Bog with that from Pineo Pond (Winne 1988), which was constrained by conventional Pb^{210} dating.

Table 1. Radiocarbon dates from South Bog, Cutler, ME (YBP = years before present).

Depth (cm)	Lab Sample number	Material - type of dating	C^{14} age, YBP	C^{13}/C^{12}	C^{13} adjusted age, YBP	Calibrated result (calendar years)
18–22	Beta-71009	peat - bulk, extended count	720 ± 40	−27.1	690 ± 40	A.D. 1270 to 1320, A.D. 1340 to 1390
33–37.5	Beta-71010	peat - bulk, extended count	2060 ± 40	−27.3	2020 ± 40	100 B.C. to A.D. 80
75–80	Beta-71011	peat - bulk	2930 ± 60	−26.8	2900 ± 60	1270 to 910 B.C.

Table 2. Changes in the grassland/forest balance of the East Stream Grassland, Cutler, ME.

Time period	Grassland that became forested	Forest that became grassland	Net expansion of grassland	Portion of grassland experiencing no change of vegetation cover
1938–1941 to 1988	20%	10.1%	−9.7%	80%
1968–1988	13%	27.2%	18.5%	87%

The remaining material was sieved through a 425-micrometer (µm) mesh to remove large charcoal particles. The total area of charcoal and size of the largest particles were noted. Pollen-slide charcoal, smaller than 143 µm, tends to be underrepresented near the site of an actual fire, and its source area can actually be subcontinental to global (Clark 1988); it was therefore disregarded. Macroscopic charcoal particles are more likely to have originated in near proximity to their point of deposition than are pollen-slide charcoal particles (Clark 1988).

Maps and Aerial Photographs

I used forest-type maps for 1938–1941, 1958, 1968, and 1978; aerial photographs from 1955, 1968, 1979, and 1988 (produced by the James Sewall Co., Old Town, ME); and a vegetation map for 1988 (prepared from aerial photographs) to compare the extent of the grassland at various times by overlaying maps of different ages. By scanning the maps into a computer-readable format, I was able to rotate, skew, and alter their sizes to bring them into the same scale.

Results

History

N. Beam (pers. comm.), a lifelong resident of Cutler, indicated that the East Stream Grassland was covered by coniferous forest until the early 1870s when an accidental fire swept the site. Following the fire, vast "bluejoint meadows" developed. These have been maintained, according to Beam, by periodic fires set by local hunters.

In 1794 Cutler was said to have been "valuable for its forests of timber, mainly spruce" (Drisko 1904). In 1835 "the whole township was covered with a heavy growth of spruce timber, with some ridges of hard wood and only here and there a pine tree" (Wilder 1879). The majority of what is now the East Stream Grassland site was purchased in 1836 by a company that produced lumber and lathe until 1850 (Washington Co. Registry of Deeds book 34, p. 450; U.S. Dept. Agric. census of industry 1850).

Regional newspapers reported prolonged drought accompanied by extensive fires during the summer of 1854 in coastal and central Maine, New Brunswick, Nova Scotia, and Vermont. On 22 August 1854, for example, the *Machias Union* reported that "in this region we appear to be surrounded by fires, which are raging in the woods and on the meadows, to an alarming extent. Large tracts of land, some very valuable for their growth of wood and timber, have been completely swept over by the devouring element."

Although Cutler was not named specifically in the newspaper reports, many sites close to it were noted to have burned that summer. According to Wilder (1879), the fires of the 1850s "completely destroyed all of the timber [in Cutler] then left by the axe." The lumber and lathe company closed its sawmills because of insufficient timber (Wilder 1879).

No farms were listed within the bounds of what is now the East Stream Grassland in the agricultural census records of 1850, 1860, or 1870, but in 1872 several individuals established farms in the vicinity. The *Machias Union* (14 May 1872) described the site as follows: "The land is excellent. The hay privilege if improved is very simple. Hundreds of tons will be cut this season." In 1880 the agricultural census reported that the owners of the site possessed 24 ha of permanent meadows and 1,821 ha of "other unimproved land" including old fields and growing wood. In 1881 a cheese factory was established just south of the confluence of Bagley Brook and East Stream, and 23 cows were pastured there (*Machias Union* 5 April and 5 July 1881).

Maine Forest Service records of fire events since 1967 reveal that 152 ha of the 800-ha East Stream Grassland have been burned since that time, with 6 ha having burned twice. The location of these fires is indicated in Figure 1.

Charcoal and Pollen Analysis

Pollen analysis revealed that birch (*Betula*), eastern hemlock (*Tsuga canadensis*), balsam fir, and white pine (*Pinus strobus*) were the predominant contributors to the pollen rain for at least 3,000 years prior to the European settlement period that began in the late 1700s (Fig. 2). Spruce (*Picea*) pollen began to be deposited in significant quantities about 1,000 years ago. The top meter of South Bog sediment did not contain large quantities of macroscopic charcoal prior to the 1300s.

Sometime during the 1300s the deposition of macroscopic charcoal began to increase, and grass (Poaceae) and aster (Tubuliflorae) type pollen began to be deposited. About 1770, ragweed (*Ambrosia*) pollen abundance increased and charcoal-particle deposition continued to increase. About 1835, grass, aster, and alder pollen abundance increased further, and a charcoal peak occurred. A second charcoal peak appeared a short time later with an increase of heath (Ericaceae) pollen abundance. Although macroscopic charcoal accumulation decreased by 1.5 to 2.5 orders of magnitude in the top three samples of the core, the pollen assemblage remained stable.

Maps and Aerial Photographs

Changes in the vegetation cover of the East Stream Grassland between 1938–1941 and 1988 and between 1968 and 1988 are shown in Table 2 and in Figures 3 and 4.

Discussion

Establishment of the Grassland

Historic evidence is consistent with local accounts that the East Stream Grassland came into existence by the early 1870s and probably following fires in the 1850s. The site was probably forested prior to 1850, as evidenced by (1) Drisko's (1904) statement that in 1794 Cutler was valuable for its forests; (2) the 1850 census record of lumber and lathe production on the site; and (3) the fact that the census records do not list farms on the site until 1880.

The closing of the lumber company's sawmills in the 1850s, the presence of large farms (including 1,821 ha of "old fields" and "growing trees") beginning in 1872, and the establishment of the cheese factory in 1881 support pre-1872 existence of the grassland.

Pollen indicators of open vegetation, such as grass and herbs, were sparse prior to about 1770 when the first European settlers arrived in the Cutler area. The increases in ragweed pollen abundance and in macroscopic charcoal indicate the arrival of European settlers. If a grassland as large as the East Stream Grassland had existed adjacent to South Bog prior to settlement, some evidence of open conditions should be apparent in the pollen stratigraphy before the European-

Paleoecology of the Cutler, Maine, Grassland

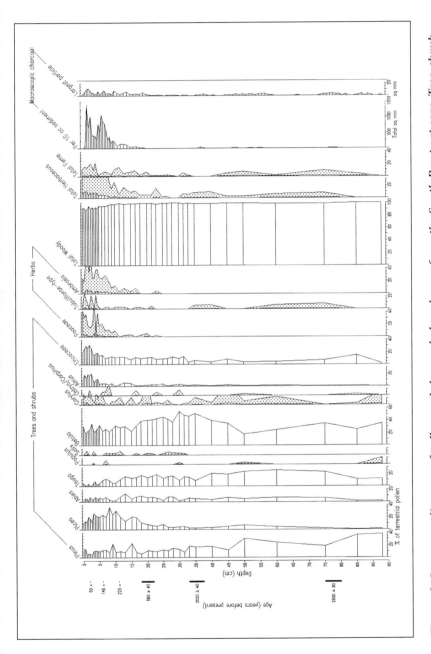

Figure 2. Summary diagram of pollen and charcoal abundances from the South Bog peat core. Tree, shrub, and herb values are percentages based on the sum of all terrestrial pollen excluding alder and Cyperaceae. Shaded curves are exaggerated by a factor of 10. Starred dates were inferred from Winne 1988. Redrawn with permission from Dieffenbacher-Krall 1996.

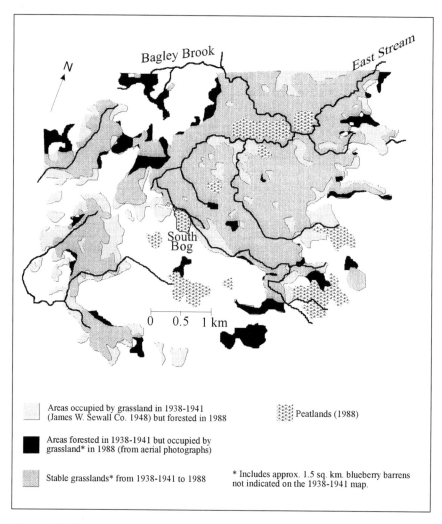

Figure 3. Changes in grassland/forest ecotone from 1938–1941 to 1988. Redrawn with permission from Dieffenbacher-Krall 1996.

settlement ragweed rise. Concurrent declines in fir and hemlock pollen abundances may have resulted from logging activities. The second charcoal peak, appearing a short time later with increased heath pollen abundance, may have been associated with increased human management of blueberry barrens near South Bog.

Role of Fire in Maintaining the Grassland

The stability of 87 percent of the area of the East Stream Grassland from 1968 to 1988 (Fig. 4), despite the fact that only 20 percent of the

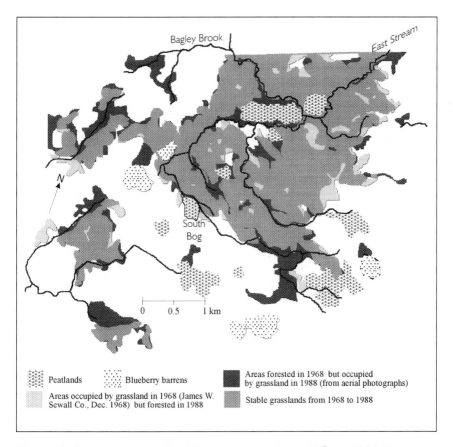

Figure 4. Changes in grassland/forest ecotone from 1968 to 1988. Redrawn with permission from Dieffenbacher-Krall 1996.

area has burned since 1967, suggests that fire may not have played the primary role in maintaining the majority of the grassland. This conclusion is supported by the stability of the pollen assemblage for the past 100 years, despite the recent decrease in macroscopic charcoal deposition to nearly zero. Shifts in the vegetative cover of substantial portions of the grassland and the surrounding forest, however, indicate that disturbance may nonetheless be important for maintaining the grassland/forest balance. Disturbances that have resulted in the woody invasion of some portions of the grassland and the creation of new areas of grassland may include small-scale logging, windthrows, and fire. Fire may have occurred more frequently on the East Stream Grassland than is documented because the Maine Forest Service (MFS) is required to report fires only when MFS personnel have taken action to control them (MFS 1993). Small fires that were controlled by landowners or were extinguished naturally may not have been included in the MFS records, but they were reported by local citizens to have occurred.

Since approximately 1940 the total area of the East Stream Grassland has declined by only 9.7 percent, although the borders have shifted in some places. However, numerous clear-cut patches outside the site on 1955 aerial photographs were being invaded by woody plants in 1968 and 1978 aerial photographs, and these patches were indistinguishable from the surrounding forest in 1988 aerial photographs. The effects of disturbance on the East Stream Grassland are unpredictable given the current incomplete history of disturbance at the site; at some times in Cutler's past, disturbance has resulted in the establishment of bluejoint reedgrass grassland, and at other times it has not.

Bluejoint Reedgrass Autecology

The current vegetation of the East Stream Grassland may itself be responsible for its continued existence. Although bluejoint reedgrass grasslands are unusual in Maine, they are widespread across western North America where almost pure stands of this species sometimes develop on both lowland and upland mesic sites (Mitchell and Evans 1966, Mitchell 1974, Smith and James 1978). The bluejoint reedgrass stands in Cutler, occurring continuously from wetland soil types to well-drained droughty soils, are consistent with this pattern. The moisture contributed by the frequent fog may allow a typically hydrophytic plant to maintain good growth and dominance on well-drained soil.

Bluejoint reedgrass may prevent reforestation by several means. In as few as three or four years, the accumulation of undecayed stems from previous years' growth can form a continuous mat of dense thatch that can prevent the seeds of other plants from reaching mineral soil before they germinate (Eis 1981, Haeussler and Coates 1986). Once germinated, the suspended seedlings desiccate before their roots reach the soil or because the roots cannot penetrate the dense mat (Haeussler and Coates 1986). Not only do tree and shrub seedlings fail to become established on such sites, but planted conifers do not succeed either (Hogg and Lieffers 1991b). The thatch in many locations of the East Stream Grassland is 10 to 20 cm thick.

The thick thatch has the additional property of insulating the soil. Hogg and Lieffers (1991a) measured soil temperatures under unmowed, frequently mowed, and once-mowed patches of bluejoint reedgrass. They found that the presence of thatch in the unmowed plots delayed the warming of soil in the spring and reduced the soil temperature during the summer by as much as 5°C. Diurnal temperatures were affected in the same manner; soil warmed more slowly in the morning in the unmowed plots than in those lacking thick thatch, and the maximum soil temperature attained during the day was reduced. Colder soil may reduce root growth, water uptake, stomatal conductance, and photosynthetic rate and may delay the timing of bud flush of conifer seedlings (Hogg and Lieffers 1991a).

Reaching heights of 1 to 1.5 m (Mitchell 1974), bluejoint reedgrass can further limit the growth of other seedlings by shading them and by smothering them as the grass's dead standing growth from the previous year is compressed by snow (Eis 1981). Bluejoint reedgrass forms a dense root network 1 to 2 cm deep and can easily outcompete tree seedlings for water, especially on well-drained or sloping sites where surface water is transient (Sims and Mueller-Dombois 1968). In addition, bluejoint reedgrass competition for nutrients has been found to have a negative effect on the growth of white spruce seedlings (Rivard et al. 1990).

Preservation of the East Stream Grassland

The East Stream Grassland is likely to continue in its current state if other conditions, including the on-site activities of local citizens, do not change. As indicated by the relative stability of the grassland since 1968 in the absence of large fires, large-scale burning is probably not necessary to maintain the site. Future monitoring is necessary to test this hypothesis. Indeed, several recent studies of bluejoint reedgrass grasslands (e.g., Blackmore and Corns 1979, Conn and Deck 1991, Hogg and Lieffers 1991b) have investigated not how to preserve this vegetation type but how to eliminate it.

Changing conditions, such as climate, could alter the current competitive balance between bluejoint reedgrass and other species. If a heavy fire that damaged the grass's rhizomes were followed by several years of growing conditions that were poor for the grass but suitable for tree and shrub seedlings, young trees and shrubs might be able to become established before the grass developed the thick thatch mat that currently seems to be suppressing tree growth. This may explain how the few even-sized stands of quaking aspen that dot the grassland became established and may explain the fluctuations in the grassland/forest ecotone illustrated in Figures 3 and 4. Warmer and drier conditions may stimulate tree and shrub growth because the hydrophytic grass may grow less vigorously and its thatch decay more rapidly.

To maintain the current balance between grassland and forest, the East Stream Grassland must be monitored regularly for extensive ecotone shifts. A ground survey every few years and aerial photography every 5 to 10 years could reveal if either vegetation type is encroaching upon the other. If grassland does appear to be decreasing significantly, test burns or cutting of woody vegetation should be conducted in small areas to determine the effects of an extensive disturbance under current climate conditions. Grazing is not recommended as a conservation tool because bluejoint reedgrass is intolerant of heavy grazing and because repeated harvests and grazing may prevent the accumulation of the thatch that suppresses tree growth (McKendrick et al. 1977).

Acknowledgments

I thank G. L. Jacobson, Jr., R. B. Davis, D. C. Smith, C. Norden, H. Almquist-Jacobson, M. Spencer-Famous, N. Famous, B. Vickery, K. Bates, C. C. Dorian, P. I. Moreno, J. D. Dieffenbacher-Krall, T. Parent, C. Hammond, N. Beam, R. Kord, J. Cates, P. G. Nilles, T. Henze, D. Cameron, P. Vickery, and P. W. Dunwiddie for invaluable suggestions, information, and field and laboratory assistance. I also thank two anonymous reviewers. Funding for this project was provided by the Maine Coast Heritage Trust.

Literature Cited

Aaby, B., and G. Digerfeldt. 1986. Sampling techniques for lakes and bogs. Pp. 181-194 *in* Handbook of Holocene paleoecology and paleohydrology (B. E. Berglund, ed.). John Wiley, New York.

Blackmore, D. G., and W. G. Corns. 1979. Lodgepole pine and white spruce establishment after glyphosate and fertilizer treatments of grassy cutover forest land. Forest Chron. 55: 102-105.

Bliss, L. C., and R. W. Wein. 1972. Plant community responses to disturbances in the western Canadian Arctic. Can. J. Bot. 50: 1097-1109.

Cates, J., and A. Dennison. 1976. A brief history of Cutler and some interesting incidents. Cutler Assoc., Cutler, ME.

Clark, J. S. 1988. Particle motion and the theory of charcoal analysis: source area, transport, deposition, and sampling. Quat. Research 30: 67-80.

Conn, J. S., and R. E. Deck. 1991. Bluejoint reedgrass (*Calamagrostis canadensis*) control with glyphosate and additives. Weed Tech. 5: 521-524.

Dieffenbacher-Krall, A. C. 1996. Paleo- and historical-ecology of the Cutler grasslands, Cutler, Maine: implications for future management. Nat. Areas J. 16: 3-13.

Drisko, G. W. 1904. Narrative of the town of Machias: the old and the new, the early and the late. Press of the Republican, Machias, ME.

Eis, S. 1981. Effect of vegetative competition on regeneration of white spruce. Can. J. Forest Res. 11: 1-8.

Famous, N., and M. Spencer. 1991. Ecological survey of the Hearst property, Cutler/Whiting, Maine. Report to Maine Coast Heritage Trust, Brunswick.

Fœgri, K., P. E. Kaland, and K. Krzywinski. 1989. Textbook of pollen analysis. John Wiley, Chichester, U.K.

Haeussler, S., and D. Coates. 1986. Autoecological characteristics of selected species that compete with conifers in British Columbia: a literature review. Br. Columbia Ministry of Forests, Victoria.

Hogg, E. H., and V. J. Lieffers. 1991a. The impact of *Calamagrostis canadensis* on soil thermal regimes after logging in northern Alberta. Can. J. Forest Res. 21: 387-394.

Hogg, E. H., and V. J. Lieffers. 1991b. The relationship between seasonal changes in rhizome carbohydrate reserves and recovery following disturbance in *Calamagrostis canadensis*. Can. J. Bot. 69: 641-646.

Jowsey, P. C. 1966. An improved peat sampler. New Phytologist 65: 245-248.

Maine Forest Service. 1993. Town warden manual, downeast district. Dept. Conservation, Machias, ME.

McKendrick, J. D., A. L. Brundage, and V. L Burton. 1977. Quality of bluejoint hay is influenced by time of harvest. Agroborealis 9: 26-29.

McMahon, J. 1992. An evaluation of the conservation potential of the Hearst property in Cutler and Whiting, Maine. Report to Maine Coast Heritage Trust, Brunswick.

Mitchell, W. W. 1974. Native bluejoint: a valuable forage and germplasm resource. Agroborealis 6: 21-22.

Mitchell, W. W. 1975. Climate reversals and Alaska's grasslands. Agroborealis 7: 12-15.

Mitchell, W. W., and J. Evans. 1966. Composition of two disclimax bluejoint stands in southcentral Alaska. J. Range Manage. 19: 65-68.

Rivard, P. G., P. M. Woodard, and R. L. Rothwell. 1990. The effect of water table depth on white spruce (*Picea glauca*) seedling growth in association with the marsh reed grass (*Calamagrostis canadensis*) on wet mineral soil. Can. J. Forest Res. 20: 1553-1558.

Sims, H. P., and D. Mueller-Dombois. 1968. Effect of grass competition and depth to water table on height growth of coniferous tree seedlings. Ecology 49: 597-603.

Smith, D. W., and T. D. W. James. 1978. Changes in the shrub and herb layers of vegetation after prescribed burning in *Populus tremuloides* woodland in southern Ontario. Can. J. Bot. 56: 1792-1797.

U.S. Department of Agriculture Soil Conservation Service. 1983. Soil survey field sheet, Washington Co., ME. U.S. Dept. Agric. Soil Conserv. Serv., Fort Worth, TX.

Wilder, I. 1879. Cutler. Special collections, Univ. of Maine Library, Orono.

Winne, J. C. 1988. The history of vegetation and fire on the Pineo Ridge blueberry barrens, Maine. Master's thesis, Univ. of Maine, Orono.

Ann C. Dieffenbacher-Krall: Department of Plant Biology and Pathology, University of Maine, Orono, ME 04469.

The Role of Nutrient-level Control in Maintaining and Restoring Lowland Heaths: British and Northern European Techniques of Potential Application in Northeastern North America

Wesley N. Tiffney, Jr.

Abstract

Heathlands in northern Europe and northeastern North America have similar origins, developmental histories, and abandonment and decline histories. European, and particularly British, research on heath ecology and management techniques, with particular emphasis on controlling nutrient levels derived from agricultural or atmospheric sources, has been ongoing for decades longer than the comparatively recent commencement of such studies in North America. Therefore, New World heath managers will profit from an understanding of Old World experience with fire, brushcutting, vegetation removal, control of resprouting by invasive vegetation, crop harvesting, grazing, deep plowing, and turf removal. Heath restoration may be accomplished by transplanting turf, spreading soil from other heathlands, or using mowing clippings as seed sources on new heaths. This paper discusses these techniques and their possible application to northeastern North American lowland heaths.

W. N. Tiffney, Jr.

Parallels between Northern European and Northeastern North American Heaths

Northern European heaths (characterized by heather [*Calluna vulgaris*]) and northeastern North American lowland heaths, best represented on Nantucket Island, MA (characterized by bearberry [*Arctostaphylos uva-ursi*]), are anthropogenic. They spread from previous small patches and were subsequently maintained by massive habitat alteration, including cutting for timber and fuel, periodic fire, turf and peat cutting, and, most importantly, widespread pasturing of domesticated grazing animals. British heaths began developing some 5,000 years ago when grazing animals were introduced (Gimingham 1972, Webb 1986). Nantucket's deforestation was influenced by 9,000 years of Native American occupation, but actual heath development accelerated when European settlers introduced domestic herbivores. British heaths peaked in extent about 1750 and those on Nantucket about a century later. On Nantucket, some 23,000 sheep-grazing equivalents were in use on the 12,500-hectare (ha) island by 1850 (Tiffney and Eveleigh 1985, Dunwiddie 1989).

Both British and Nantucket heaths had largely disappeared by 1900 because of the decline of traditional land-use practices. In Britain, changes in land use began as early as the late 1700s, as fertilizers and improved farm machinery made it possible to convert heaths to more productive arable or forestry lands (N. R. Webb pers. comm.). Later decreases in the extent of heath habitat resulted from a decline in grazing and from poor management, including both lack of prescribed fire and inappropriate burning techniques (Webb and Haskins 1980, Farrell 1989, Webb 1990). On Nantucket, grazing animals disappeared rapidly after 1900, fossil fuels replaced wood as a source of fuel, and universal fire suppression became the rule (Tiffney and Eveleigh 1985). British and Nantucket heaths also lost area to urban and suburban development. Today some 10 to 15 percent of original heaths remain in most of Britain, and smaller amounts survive elsewhere in northern Europe. The same is true on Nantucket, and only isolated small patches survive on Martha's Vineyard and Cape Cod, MA, and elsewhere in northeastern North America, primarily at coastal sites.

In both Britain and northeastern North America, rare plants and animals dependent on heath habitat are presently endangered or have been lost. Of particular concern is the process of heath fragmentation, in which edge succession by brush or trees results in progressively smaller heath patches. Genetic exchange among plants and animals restricted to these increasingly small and separated heath "islands" becomes more difficult, thus imposing greater stress on the survival of heath-dependent species.

Present management objectives for British and Nantucket heaths are essentially the same: managers seek to halt further loss of habitat, to rejoin fragmented heaths, and in some cases to reclaim former heaths lost to brush and tree succession. British managers have been working toward these objectives for several years, whereas Nantucket management efforts are just beginning on a large scale.

High Nutrient Levels and Heath Loss

In general, dwarf shrub heaths are characteristic of nutrient-poor sites (Gimingham 1972). As nutrients on heaths increase, either from intentional application or atmospheric sources, so does the probability that plants better adapted to higher nutrient levels will invade and outcompete characteristic heath plants such as heather or bearberry. In some cases, increased nutrients may simply be a consequence of shrub and tree succession into heaths, as in the case of scrub oak (*Quercus ilicifolia*) invasion on Nantucket heaths. Increased litter produced by the larger biomass of shrubs and trees results in a more nutrient-rich and nutrient-retentive ground layer of organic material deposited over the original glacial sands and gravels.

In Europe, an indication that increased nutrient levels may be fostering succession to nonheath forms may be heralded by the spread of hairgrass (*Deschampsia flexuosa*) and purple moor grass (*Molinia caerulea*). As hairgrass also occurs in some northeastern North American heaths, notably on Nantucket, its spread may warn of increasing nutrient levels here as well (Hester et al. 1991, Gimingham 1992).

Although high nutrient levels may be encountered by managers seeking to reestablish heaths on former croplands, much current concern is directed toward rising nutrient levels associated with acid precipitation. In Britain, the lowest wet atmospheric nitrogen deposition levels occur near coasts and may average from below 10 to 10–15 kilograms/hectare/year (kg/ha/yr). Over much of inland Britain, levels within the range of historic or surviving heaths range from 15 to 25 kg/ha/yr, and levels above 30 kg/ha/yr have been measured (Bell 1994). In the Netherlands, atmospheric input of nitrogen to the few surviving Dutch heaths has been measured at levels of 40 to 80 kg/ha/yr (de Smidt and van Ree 1992). In the northeastern United States, wet nitrogen input rates average 9 to 11 kg/ha/yr, and total (wet and dry) nitrogen deposition rates of 10 to 13 kg/ha/yr have been recorded for the higher elevations of New York, Vermont, and New Hampshire. In Massachusetts in the coastal area of Cape Cod and on the islands of Martha's Vineyard and Nantucket, where most of northeastern North America's surviving heaths are located, wet rates are 3 to 5 kg/ha/yr, and total rates average 6 to 8 kg/ha/yr (Ollinger et al. 1993).

Hence, although nutrient deposition from atmospheric sources in Britain and Europe is much higher than in northeastern North America, cumulative deposition in North America may still be an important factor in heath succession to scrub and brush cover. This may be the case when nutrient-retaining litter layers are allowed to accumulate over previously permeable and easily leached heath soils. North American managers may wish to take advantage of comparatively low atmospheric nutrient inputs to actively maintain low nutrient levels within surviving heaths by applying a series of British and European nutrient-control management techniques.

British and European Heath Management Techniques, with Emphasis on Maintaining Low Nutrient Levels

Fire

Prescribed fire is slowly gaining more acceptance as a management technique throughout northeastern North America generally (Patterson et al. 1985) and on Nantucket specifically (Dunwiddie 1990). This practice is still in its infancy in North America, however, and significant heath tracts have yet to be burned. In Scotland the general body of evidence suggests that nutrient losses from burning (smoke and fly ash) may approximate that from atmospheric input, although phosphorus may be an exception (Muirburn Working Party 1977). Chapman (1967), working on southern English heaths, suggested that burning may result in both phosphorus and nitrogen depletion. Dudley (1992), working in grassy heaths on Martha's Vineyard and Nantucket, concluded that long-term depletion of nitrogen levels followed prescribed burning.

Hence, burning may help to decrease nutrient levels, but this decrease may be trivial if atmospheric nutrient contributions are substantial. Observations on Nantucket appear to support the idea that burning may maintain an existing heath against shrub and tree succession but that it does little to restore overgrown heaths. More drastic methods of nutrient reduction appear necessary for heathland restoration.

Brushcutting

Brushcutting, using a tractor-drawn or mounted mechanical cutter, is a widely used heath management technique both in Britain and on Cape Cod, Nantucket, and Martha's Vineyard. It is preferred when burning is too dangerous, when firebreaks need to be constructed, or when a clear-cut of vegetation is desired rather than the mosaic pattern left by fire. However, treatment of the cut debris, and thus its impact on nutrient levels, varies widely in Britain and Massachusetts.

In Britain, standard practice involves blowing the clippings into a silage trailer attached to the cutting machine, or otherwise collecting the clippings for export from the site. This direct nutrient export is similar to cropping a site or to pasturing grazing animals on it and then removing them. In Britain, such cuttings may be used for biofiltration purposes, mulches, foundations for roads, or seed sources for heathland restoration. Under certain circumstances for which nutrient accumulation is not a major concern, fine or sparse clippings may be left on the site, but export is the general rule (Gimingham 1992).

On Nantucket it is standard practice to leave cuttings on the site. This may result in stifling regeneration of comparatively delicate heath plants and may actually promote robust resprouting of the primary Nantucket heath invader, scrub oak. The thick layer of chipped organic debris may help retain atmospheric nutrient inputs rather than allow them to leach through normally porous sandy soils. Decomposition of the organic litter layer may also add additional nutrients.

Brushcutting as presently practiced on Nantucket may be counterproductive to heath maintenance and regeneration. Comparative studies of effective heath regrowth should be initiated in areas where clippings remain on the site and where they are removed.

Direct Vegetation Removal

Perhaps the simplest method of heath restoration may be practiced where invasive or planted brush or trees dominate a former heath but where heaths still persist as an understory and the trees and shrubs can be removed relatively easily. In Dorset, in southeast England, unwanted pine (*Pinus*) trees from former pine plantations were cut and the leaf litter raked up and removed. Heath regenerated from surviving understory plants and from a heather seed bank persistent in the soil for some 70 years (Webb and Pywell 1993). The success of this approach presupposes that removed shrub or tree vegetation will not resprout, or that resprouting will be controlled.

The Resprouting Problem

In Britain, two birch species, silver birch (*Betula pendula*) and downy birch (*B. pubescens*), are active invaders of heaths and are difficult to control as they resprout prolifically from basal buds (Marrs 1984, Gimingham 1992). On Nantucket, heath managers have encountered the same problem with scrub oak. The simplest means of controlling a persistent resprouting species is "grubbing out," or complete mechanical removal of the plant, roots and all, by hand or tractor. This is labor intensive, is quite expensive, and may lead to unacceptable soil disturbance (Gimingham 1992). Disposal of the large quantities of brush generated by this process may also be problematic. Successive cutting of resprouting species over several growing seasons may finally exhaust energy stored in the roots, but the practice is expensive, and machines used for this purpose may have a severe impact on vegetation and soil (Marrs 1984). However, mechanical removal or successive recutting (if the cuttings are removed from the site) will promote nutrient export from the heath habitat and may be the treatment of choice over the long term.

Direct foliar spraying of herbicides has been used as a control technique in Britain. One study suggested that fosamine ammonium and 2,4,5-T were the most satisfactory sprays, showing the least collateral damage to heath plants (Marrs 1984). However, managers frequently encounter reluctance to permit herbicide spraying in nature reserves.

Cutting followed by applying herbicide to cut stumps, usually with a paint brush, provides a more specific treatment of target species. Scoring or notching the stump helps retain the herbicide. Glyphosate (marketed under the trade name Roundup in the United States) has provided good results in Britain, and when carefully applied it resulted in nearly 100 percent birch kill (Gimingham 1992). In dense stands of invasive species, the shrubs may be cut and then allowed to resprout for no more that two growing seasons. This results in an even height of resprouting shoots, some distance above surviving heath species. The resprouts may then be treated by foliar spray or by applying herbicide with a wicklike weed wiper (Gimingham 1992). If herbicides are to be used, field trials should be conducted in specific areas to determine the minimum effective dose, and great care should be taken to select an application method that will minimize collateral effects on nontarget species. Herbicide applications have little effect in reducing nutrient levels.

Nutrient Reduction Techniques: Restoring Heaths on Former Croplands

Former croplands or lands that have been fertilized or limed maintain persistently high nutrient levels. Although natural leaching and erosion reduce levels of nitrogen and phosphorous after about 10 years, calcium from lime is persistent and may hold soil fertility at high levels (Pywell et al. 1994). Actively reducing nutrient levels may be necessary before reintroducing heaths to such sites.

One approach is simply to continue growing and harvesting crops from former agricultural sites without adding additional fertilizer. Harvesting and removing a crop, such as hay, constitute direct nutrient export (Pywell et al. 1995).

Grazing animals may be used to accomplish the same objective if they are removed from the site at night (Gimingham and de Smidt 1983). Called "rotational grazing" today and regarded by some as a new approach, this is a variation on an ancient technique that was originally used to fertilize a fallow field rather than to reduce nutrients. In the eighteenth century, J. Hector St. John de Crevecoeur (*in* Philbrick 1994) observed what Nantucket shepherds did when they folded their sheep in a field they wished to fertilize: "Three times during a night it is permissible to terrify their sheep with burning coals; each time terror forces them to depose their manure; during one night a flock of these animals fertilize and enrich to a great extent the field in which they are enclosed." Quietly reversing this procedure reduces nutrient deposition.

Grazing techniques must be applied with care. Undergrazing in British heaths may allow overmature "woody" heather to develop, and gaps in senescent plants may allow succession by undesired species. Overgrazing can damage heather, leading to dieback and replacement by undesired grasses. When carefully applied, however, controlled grazing presents a useful management option throughout Britain (Gimingham 1992). In Denmark, heath maintenance and brush-succession control have been successfully accomplished using primitive cattle breeds, particularly belted Galloway (Buttenschon and Buttenschon 1982, 1985, pers. comm.).

Although Nantucket and other northeastern North American lowland heaths were produced by sheep and cattle grazing, there is no modern experience with using these animals to maintain or restore heaths. Sheep grazing decreased on Nantucket heaths after about 1900 (Tiffney and Eveleigh 1985). Today the expenses of maintaining grazing animals (including veterinary care, supplementary food, and winter shelter) plus problems with the public and unrestrained dogs preclude using sheep in large-scale programs of heath maintenance and restoration on Nantucket. However, the possibility exists that primitive cattle breeds, able to defend themselves against aggressive people and dogs, may be of limited use. This should be assessed experimentally.

In extreme cases of nutrient saturation in Britain and the Netherlands, deep plowing or turf removal has been used to lower nutrient levels to the point where heaths can again survive. The principle involved in plowing is to mix nutrient-rich upper soil layers with nutrient-poor mineral layers below, thus reducing total nutrients available in the rooting zone. Sterile materials such as sand or quarry rubble may be added to enhance dilution. To be effective, turnover depth should be at about 24 centimeters (cm), or below the A horizon (the dark-colored horizon with high organic content and minerals).

Simple rototilling apparently is not deep enough to have a significant effect (Pywell et al. 1995).

Turf removal, involving removal of nutrient-saturated upper soil levels, has been used successfully in extreme cases of nutrient contamination. Again, removal depth should be below the A horizon, at about 24 cm depth. Off-site disposal of contaminated soil is usually necessary (Smith et al. 1991, Pywell et al. 1995).

Both plowing and turf removal may be successfully applied when the nutrient source is previous fertilization on abandoned farmland. If contaminating nutrients were derived from atmospheric sources, however, these techniques at best will constitute a temporary solution unless atmospheric inputs have been controlled.

Reestablishing Heath Vegetation

If there are no surviving heath plants on the site and no seeds remaining in the soil, reintroduction of heath vegetation will be necessary. This may be accomplished in several ways.

The most effective method is to translocate turf from existing heaths. This method is expensive and has the undesirable effect of damaging and reducing other heathlands, but it may be justified if the new heath is small and will serve as a source for propagule dissemination. A more efficient use of other heaths as propagule sources is to spread heath topsoil on the site to be regenerated. In Britain this resulted in rapid regeneration and allowed three to five times the source heath to be treated. A third technique is to apply clippings containing heath-plant seeds to the restoration area. In Britain this was not as effective or as fast-acting a technique as planting turf or spreading topsoil, but it did result in regeneration with minimum impact on the source heathland (Pywell et al. 1995).

Although turf- and soil-transplanting techniques should work well on northeastern North American heaths, the heath species here do not produce seeds with the abundance of the dominant British heath plant, heather. Also, bearberry, a notably recumbent shrub, would likely yield few fruits if it was clipped and the clippings were collected. Direct transplant of turf or plants may prove the most effective approach.

Summary

This paper describes several heath management and restoration techniques now used in Britain and northern Europe. Some or all of these may be useful to northeastern North American heath managers. Most of these techniques may be seen in active application in Dorset, England, where the Royal Society for the Protection of Birds, English

Nature, and other organizations have embarked on a massive heathland restoration program. Ultimately, British restoration workers seek to create 6,000 ha of heath habitat nationally. This demonstrates the importance that northern European and British authorities have assigned to restoring, regenerating, and maintaining heaths (Auld et al. 1992, M. Auld and N. Webb pers. comm.).

Acknowledgments

My sincere and abundant thanks are due to British colleagues M. Auld, S. B. Chapman, L. Farrell, and particularly C. H. Gimingham and N. R. Webb for the many hours they have spent in the field and in discussion of British and northern European heath ecology. I acknowledge my continuing debt to the late J. C. Andrews who led me to understand so much about Nantucket and its land-use history. P. W. Dunwiddie reviewed and improved this manuscript.

Literature Cited

Auld, M., S. Davies, and B. Pickess. 1992. Restoration of lowland heaths in Dorset. Royal Soc. Protect. Birds Conserv. Rep. 6: 68-73.

Bell, N., ed. 1994. The ecological effects of increased aerial deposition of nitrogen. Br. Ecol. Soc. Ecol. Issues 5, Field Studies Council, Montford Bridge, Shrewsbury Bridge, U.K.

Buttenschon, J., and R. M. Buttenschon. 1982. Grazing experiments with cattle and sheep on nutrient poor, acidic grassland and heath. I: vegetation development. Nat. Jutlandica vol. 21 (no. 1-3): 1-48.

Buttenschon, J., and R. M. Buttenschon. 1985. Grazing experiments with cattle and sheep on nutrient poor, acidic grassland and heath. IV: establishment of woody species. Nat. Jutlandica vol. 21 (no. 7): 117-140.

Chapman, S. B. 1967. Nutrient budgets for a dry heath ecosystem in the south of England. J. Ecol. 55: 677-689.

de Smidt, J. T., and P. J. van Ree. 1992. The loss of cryptograms in Dutch heathland as an effect of the atmospheric deposition of nitrogen. Pp. 29-4 to 29-6 *in* Fourth international European heathland workshop (B. Clement, ed.). Université de Rennes, Bretagne, France.

Dudley, J. L. 1992. Secondary succession and nitrogen availability in coastal heathlands. Ph.D. diss., Boston Univ., Boston, MA.

Dunwiddie, P. W. 1989. Forest and heath; the shaping of vegetation on Nantucket Island. J. Forest Hist. 33: 126.

Dunwiddie, P. W. 1990. Rare plants in coastal heathlands: observations on *Corema conradii* (Empetraceae) and *Helianthemum dumosum* (Cistaceae). Rhodora 92: 22-26.

Farrell, L. 1989. The different types and importance of British heaths. Bot. J. Linnean Soc. London 101: 291-299.

Gimingham, C. H. 1972. Ecology of heathlands. Chapman and Hall, London.

Gimingham, C. H. 1992. The lowland heathland management handbook. English Nature Sci. no. 8, Northminster House, Peterborough, U.K.

Gimingham, C. H., and J. de Smidt. 1983. Heaths as natural and semi-natural vegetation. Pp. 185-199 *in* Man's impact on vegetation (M. J. A. Werger and I. Ikusima, eds.). Junk, The Hague, Netherlands.

Hester, A. J., J. Miles, and C. H. Gimingham. 1991. Succession from heather moorland to birch woodland. I. Experimental alteration of specific environmental conditions in the field. J. Ecol. 79: 303-315.

Marrs, R. H. 1984. Birch control on lowland heaths: mechanical control and the application of selective herbicides by foliar spray. J. Applied Ecol. 21: 703-716.

Muirburn Working Party. 1977. A guide to good muirburn practice. Dept. Agric. and Fish. for Scotland, Nature Conservancy Council, Her Majesty's Stationery Office, Edinburgh, U.K.

Ollinger, S. V., J. D. Aber, G. M. Lovett, S. E. Millham, R. G. Lathrop, and J. M. Ellis. 1993. A spatial model of atmospheric deposition for the northeastern U.S. J. Ecol. Appl. 3: 459-472.

Patterson, W. A., III, K. E. Saunders, L. J. Horton, and M. K. Foley. 1985. Fire management options for coastal New England forests: Acadia National Park and Cape Cod National Seashore. Pp. 360-365 *in* Proceedings: symposium and workshop on wilderness fire, Missoula, Montana, November 15-18, 1983. Gen. Tech. Rep. INT-182, U.S. Forest Serv., Ogden, MT.

Philbrick, N. 1994. Away off shore, Nantucket Island and its people, 1602-1890. Mill Hill Press, Nantucket, MA.

Pywell, R. F., N. R. Webb, and P. D. Putwain. 1994. Soil fertility and its implications for the restoration of heathland on farmland in southern Britain. Biol. Conserv. 70: 169-181.

Pywell, R. F., N. R. Webb, and P. D. Putwain. 1995. A comparison of techniques for restoring heathland on abandoned farmland. J. Applied Ecol. 32: 400-411.

Smith, R. E. N., N. R. Webb, and R. T. Clarke. 1991. The establishment of heathland on old fields in Dorset, England. Biol. Conserv. 57: 221-234.

Tiffney, W. N., Jr., and D. E. Eveleigh. 1985. Nantucket's endangered maritime heaths. Pp. 1093-1109 *in* Coastal zone '85 (O. T. Magoon, H. Converse, D. Miner, D. Clark, and L. T. Tobin, eds.). Am. Soc. Civil Engineers, New York.

Webb, N. R. 1986. Heathlands. Collins, London.

Webb, N. R. 1990. Changes on the heathlands of Dorset between 1978 and 1987. Biol. Conserv. 47: 153-165.

Webb, N. R., and L. E. Haskins. 1980. An ecological survey of the heathlands in the Poole Basin, Dorset, England in 1978. Biol. Conserv. 17: 281-296.

Webb, N. R., and R. F. Pywell. 1993. Heathland restoration: the potential of old fields. Pp. 48-60 *in* Proceedings of the seminar on heathland habitat creation (T. Free and M. T. Kitson, eds.). Sizewell Power Station, Suffolk, U.K.

Wesley N. Tiffney, Jr.: University of Massachusetts Nantucket Field Station, 180 Polpis Road, Nantucket, MA 02554.

Coming Full Circle: Restoring Sandplain Grassland Communities in the State Forest on Martha's Vineyard, Massachusetts

William H. Rivers

Early History

The 1,780-hectare (ha) Manuel F. Correllus State Forest is the largest tract of contiguous public open space on the island of Martha's Vineyard, MA (Fig. 1). Public land was first purchased on the island in 1908 by what is now the Massachusetts Division of Fisheries and Wildlife. The purpose of this 248-ha acquisition was to provide a refuge for the Heath Hen (*Tympanuchus cupido cupido*). This endemic subspecies of the Greater Prairie-Chicken was restricted to northeastern North America but had been extirpated from the mainland, and its only remnant population at the time remained on Martha's Vineyard. Predators, extreme weather, uncontrolled wildfires, and limited habitat kept the population from ever becoming self-sustaining. The last Heath Hen was seen on Martha's Vineyard on 11 March 1932.

Between 1908 and 1926, lands abutting the Heath Hen reservation were acquired by what is now the Massachusetts Department of Environmental Management (DEM). By 1926 the state forest was 1,820 ha in size.

Conservation efforts in the early twentieth century were largely centered around the rehabilitation and protection of hundreds of thousands of hectares of abandoned agricultural and cutover lands throughout New England. Prior to this period, wildfires burned nearly

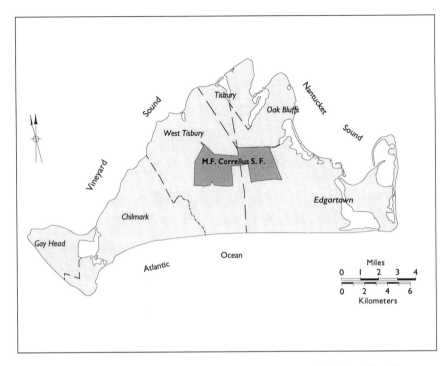

Figure 1. The island of Martha's Vineyard and Manuel F. Correllus State Forest.

40,000 ha each year in Massachusetts. Consequently, early management efforts by state-forest crews, both on Martha's Vineyard and elsewhere, consisted of establishing conifer plantations and developing a system of "fire-stops," as they were called, to protect them. The Civilian Conservation Corps planted more than 160 ha of conifers in 1934. By the end of the 1930s, approximately 400 ha in the state forest had been planted to conifers.

In 1942 the U.S. Navy took 260 ha from the center of the state forest to create the Martha's Vineyard Naval Air Station (MVNAS). Today that property is managed by Dukes County as the Martha's Vineyard Airport.

Recent Developments

Through the 1950s and 1960s, a limited amount of planting was done in the state forest. By the end of the 1960s, approximately 450 ha of the forest (25 percent) had been planted to conifers; more than 280 ha were planted to red pine (*Pinus resinosa*), and white pine (*P.*

strobus) covered approximately 80 ha. Collectively, plantings of Scotch pine (*P. sylvestris*), jack pine (*P. banksiana*), hybrid pitch pine (*P. rigida*)/loblolly pine (*P. taeda*), and white spruce (*Picea glauca*) covered another 80 ha.

In 1974 a 17.7-kilometer (km)-long bicycle trail was built in the state forest, primarily along the perimeter. This is the only major recreation facility in the state forest. It is estimated that there are nearly 100,000 annual visitors on this trail.

Ecological Considerations

After three-quarters of a century of nurturing, it is apparent that the state forest will never be what its founders had hoped. With each passing storm, the stands of planted trees become more and more bedraggled. Insects and fungi slowly take their toll. Native plant species fill the voids. In less than a century, the state forest has come full circle.

Since the 1970s, the plantings have been decimated by weather and disease. Red pine has been particularly hard hit by a fungus, *Diplodia pinea*, that apparently thrives in the maritime environment. Efforts to salvage the diseased stands have been ongoing for almost 20 years, although they have been hampered by inadequate island markets.

In August 1991, Hurricane Bob struck coastal New England. More than 80 ha of plantations and natural stands were damaged. A microcomputer simulation of fire behavior in these stands showed that increased fuel loadings caused potential flame lengths in the most seriously damaged stands to increase as much tenfold, to more than 3 meters (m). Clearly, the most urgent task was to reduce the risk of wildfire to abutting residential properties. The Massachusetts DEM sought and received financial assistance from the Federal Emergency Management Agency for hazard-reduction work in the damaged stands and for rehabilitation of perimeter firebreaks.

Concurrently, the Massachusetts DEM convened a technical advisory committee to develop a plan to manage the forest. Committee membership includes the DEM, Massachusetts Audubon Society, Massachusetts Natural Heritage Program, Nantucket Conservation Foundation, The Nature Conservancy, The Trustees of Reservations, University of Massachusetts Department of Forestry and Wildlife Management, U.S. Fish and Wildlife Service, and U.S. Forest Service. Through the committee's efforts, a great deal has been learned about the forest, including the following.

- Scrub oak (*Quercus ilicifolia*) is the most widespread plant in the forest. It is present on almost two-thirds of the forest's area both as an overstory (35 percent) and understory (29 percent) component. The mixed-oak association (black oak [*Q. velutin*a], scarlet oak [*Q. coccinea*], white oak [*Q. alba*]) dominates the overstory on 35 percent of the forest.

- Although pitch pine occurs as scattered stems throughout the forest, it forms a dominant overstory association on only 3 percent of the forest's area.
- Exclusive of the 54 km of firebreaks, grassland and heathland vegetation make up only 2 percent of the forest's area.
- Several rare plant and animal species are found in the forest. Many of these are disturbance-dependent organisms found on the firebreaks (Table 1).

Several recommendations have been made in the management plan. Some of the recommendations concerned with public-safety issues, such as hazard reduction and firebreak rehabilitation, have been accomplished or are in the process of being implemented. Some of the major recommendations that have already been implemented include the following.

- As of April 1996, 6 km of perimeter firebreaks had been widened to 65 m, and 9.6 km of interior firebreaks had been widened to 30 m. This will facilitate the control of wildfires and the use of prescribed fire for ecosystem management.
- An ecosystem management strategy for the restoration and maintenance of native and rare plant communities and rare plant and animal species has been developed. Ultimately, this will result in almost all conifer plantations being converted to grassland. The relative proportion of scrub oak will remain the same but will be managed with periodic prescribed burning to provide a succession of age classes. This will have the added benefit of limiting excessive fuel buildup and the subsequent risk of fire to abutting residential properties. Much of the oak forest will be converted to a savanna association and ultimately will be maintained by prescribed burning.
- The state forest's staff now has the ability to achieve a higher level of infrastructure maintenance due to the acquisition of more suitable equipment. In 1995 a 60-horsepower, four-wheel-drive tractor and a full complement of attachments were purchased for the forest. A 5,443-kilogram (kg) Rome disk harrow was also acquired; this unit will be deployed with a large rented bulldozer to widen additional firebreaks and to do conversion work.

Table 1. Rare species occurrences on the Manuel F. Correllus State Forest, Martha's Vineyard, MA.

Common Name	Scientific Name	Mass. Natural Heritage Rank[1]	Last Observed	Community Type
sandplain agalinis	Agalinis acuta	E	1940	Grassland
Nantucket shadbush	Amelanchier nantucketensis	SC	1986	Heathland
purple needlegrass	Aristida purpurascens var. caroliniana	T	1985	Grassland
eastern silvery aster	Aster concolor	E	1917	Grassland
bushy rockrose	Helianthemum dumosum	SC	1980	Grassland
hairy wild lettuce	Lactuca hirsuta var. sanguinea	WL	1989	Scrub oak
sandplain flax	Linum intercursum	SC	1989	Grassland
Nuttall's milkwort	Polygala nuttallii	WL	1981	Grassland
post oak	Quercus stellata	WL	1994	Scrub oak
papillose nutsedge	Scleria pauciflora var. caroliniana	E	1994	Grassland
sandplain blue-eyed grass	Sisyrinchium arenicola (= fuscatum)	SC	1988	Grassland
grass-leaved ladies' tresses	Spiranthes vernalis	SC	1982	Grassland
Short-eared Owl	Asio flammeus	E	1959	Heathland
coastal heathland cutworm	Abagrotis crumbi benjamini	SC	1992	Grassland
barrens daggermoth	Acronicta albarufa	T	1992	Scrub oak
Gerhard's underwing	Catocala herodias gerhardi	T	1993	Pitch pine
Melsheimer's sack-bearer	Cicinnus melsheimeri	T	1993	Scrub oak
imperial moth	Eacles imperialis	SC	1992	Pitch pine
coastal barrens buckmoth	Hemileuca maia maia	T	1989	Scrub oak
pine barrens zale	Zale sp.	SC	1993	Scrub oak

[1]Mass. Natural Heritage Program Rank: E = Endangered, T = Threatened, SC = Special Concern, WL = Watch List

The plan also includes restoration of the headquarters complex to provide office space, equipment storage, and a visitors' facility.

The plan is now in its final draft stage and is being refined by the Technical Advisory Committee.

William H. Rivers: Bureau of Forestry, Division of Forests and Parks, Massachusetts Department of Environmental Management, Amherst, MA 01004.

Vegetation Management in Coastal Grasslands on Nantucket Island, Massachusetts: Effects of Burning and Mowing from 1982 to 1993

Peter W. Dunwiddie, William A. Patterson III, James L. Rudnicky, and Robert E. Zaremba

Abstract

In 1983 four 0.25-hectare (ha) plots were established in Ram Pasture, Nantucket Island, MA, to evaluate methods of maintaining coastal sandplain grasslands. Cover and frequency of vascular plants were measured in the plots prior to treatments and annually thereafter. One plot was burned in April, one was burned in August, one was mowed in August, and the fourth was left as a control plot. Treatments were applied biennially. Dominant species in all plots included little bluestem (*Schizachyrium scoparium*), black huckleberry (*Gaylussacia baccata*), sheep fescue (*Festuca ovina*), and swamp dewberry (*Rubus hispidus*).

Gradual changes occurred in the vegetation of all four plots during the 11-year study. Dominant vegetation in the untreated control plot shifted from grasses and forbs to heathland shrubs. This contrasted with the plots that were burned and mowed in August, in which the frequency of herbaceous species increased 62 percent (burned plot) and 83 percent (mowed plot) and the frequency and/or cover of shrubs declined. The greatest reduction among the shrubs in these two plots occurred with bayberry (*Myrica pensylvanica*), which declined eightfold in frequency in both plots. The plot that was burned in April showed similar but less pronounced effects; frequency of nonwoody species increased 40 percent, whereas cover and frequency of most shrubs remained the same. With the biennial burns examined in this study, the season of treatment influenced vegetation more than fire intensity, which was greatest in the spring.

Introduction

The sandplain grasslands and heathlands of Cape Cod, Nantucket, and Martha's Vineyard, MA, are globally endangered communities that include a diversity of rare plants, birds, and invertebrates (Dunwiddie et al. 1993). Rare species such as bushy rockrose (*Helianthemum dumosum*), sandplain flax (*Linum intercursum*), sandplain blue-eyed grass (*Sisyrinchium arenicola* [= *fuscatum*]), sandplain agalinis (*Agalinis acuta*), eastern silvery aster (*Aster concolor*), Short-eared Owl (*Asio flammeus*), Northern Harrier (*Circus cyaneus*), Upland Sandpiper (*Bartramia longicauda*), Grasshopper Sparrow (*Ammodramus savannarum*), and regal fritillary (*Speyeria idalia*) are important management concerns in some areas. These communities occur primarily where disturbances such as fire, salt spray, or grazing have limited tree growth. Fire suppression, plant succession, real-estate development, and the decline of agriculture have combined to greatly reduce the extent of these communities since the nineteenth century.

This study was conducted on Nantucket Island at Ram Pasture, a protected sandplain grassland belonging to the Nantucket Conservation Foundation, as part of a long-term Massachusetts Audubon Society program to maintain and restore coastal grasslands, heathlands, barrens, and other rare communities. Various management methods, including combinations of burning, mowing, applying herbicides, and grazing, are being studied by the Massachusetts Audubon Society in an effort to slow or reverse the encroachment of taller woody plants and to encourage the growth of herbaceous species. Several studies at Ram Pasture have assessed the impacts of burning and mowing regimes on native flora and fauna (Dunwiddie and Caljouw 1990, Dunwiddie 1991, Dunwiddie et al. 1995). Here we report changes that took place in the vegetation at Ram Pasture over the first 11 years of the study.

Study Design and Analyses

We established four 50-x-50-meter (m) square plots, separated by 10-m-wide mowed strips, in 1982 and 1983 within a several-hundred-hectare sandplain grassland. We measured the percent cover for each plant species using a line-intercept method, making measurements annually in 30 permanently marked, 1-m-long line-segments in each plot. We also calculated percent frequency of occurrence for each species based on the number of line-segments in which a species was found. Height was measured for the tallest individual of each species with more than 5 percent cover. To avoid phenological differences, we made measurements each year at the same time in July or August in each plot. Two plots (April burn and control) could not be monitored in 1991 due to salt-spray defoliation caused by an August hurricane that year. One plot was reserved as an untreated control. In the other three, we applied treatments biennially, beginning in 1983: burning in April, burning in August, and mowing in August.

In the April burn plot we established a more intensive methodology in which we monitored 150, rather than 30, line-segments in 1982. As in the other 3 plots, only 30 of these segments were remonitored annually. In 1994, all 150 segments were remonitored to compare with the original pretreatment measurements.

A detrended correspondence analysis was performed on the annual frequency data from each plot to identify directional trends in composition over time (Gauch 1982).

Results

The composition of the study plots was typical of coastal sandplains, which are dominated by graminoids and low shrubs (Table 1; Dunwiddie et al 1993). The vegetation changes that occurred in the untreated control plot over 11 years reflected the general pattern of shrub encroachment into many of the grasslands on Nantucket. Shrub cover increased in this plot from 39 (±32 SD) to 65 (±38) percent as graminoid cover declined from 70 (±27) to 27 (±17) percent. Similarly, total shrub frequency increased from 184 to 257 percent.

Graminoids, forbs, and woody plants exhibited different responses to burning and mowing. The increase in graminoids and forbs in all three treatment plots was most evident in the frequency data. Increases in these herbaceous taxa were most pronounced in the plots that were burned or mowed in August, similar but less marked in the April burn plot, and essentially unchanged in the control plot (Fig. 1). Graminoid frequency (Fig. 2) declined in all plots but declined most in the control. In contrast, forb frequency remained unchanged in all three treatment plots but declined in the control. Woody vegetation was reduced most in the two August treatment plots, where the greatest declines in shrub cover and frequency were measured. Burning in April achieved similar but less pronounced results.

Another way to visualize the different trends in the vegetation in each plot over time is to display the positions of the data from each plot on the principal axes of a detrended correspondence analysis of the annual frequency data (Fig. 3). Interpretation of these data is facilitated by plotting the locations of those species that were most heavily weighted in the analysis. The successive locations of each plot over time are connected with lines, revealing the tendency of plots to

Table 1. Percent cover (± SD) of the major species in plots in Ram Pasture, Nantucket I., MA. The 1982 and 1983 data are pretreatment. All plots were treated in odd years, beginning in 1983. The April burn plot received six treatments, whereas the August plots received five.

Species	Control		August Burn			August Mow			April Burn	
	1983	1993	1983	1993	1983	1993	1983	1993	1982	1993
agrostis (*Agrostis* spp.)	1.0 (2.7)	<1.0 (0.5)	2.0 (14.1)	<1.0 (0.5)	1.0 (2.6)	<1.0 (0.4)	2.0 (5.1)	<1.0		
bearberry (*Arctostaphylos uva-ursi*)	0.0	0.0	5.0 (16.4)	2.0 (7.0)	0.0	<1.0	<1.0	<1.0		
purple chokeberry (*Photinia floribunda* [= *Aronia arbutifolia*])	1.0 (19.8)	<1.0 (2.1)	2.0 (17.7)	<1.0 (2.1)	3.0 (11.8)	2.0 (6.8)	4.0 (20.7)	1.0 (20.6)		
bushy aster (*Aster dumosus*)	7.0 (8.3)	2.0 (2.5)	2.0 (3.8)	1.0 (1.0)	4.0 (8.0)	4.0 (3.7)	3.0 (6.5)	2.0 (1.4)		
small flat-topped aster (*Aster paternus*)	<1.0 (3.5)	<1.0 (1.4)	12.0 (15.4)	13.0 (16.2)	<1.0 (4.6)	2.0 (7.2)	2.0 (15.9)	5.0 (11.5)		
sedge (*Carex pensylvanica/umbellata*)	15.0 (15.8)	7.0 (9.5)	2.0 (3.7)	2.0 (1.9)	2.0 (6.1)	3.0 (4.0)	2.0 (5.6)	2.0 (2.2)		
sheep fescue (*Festuca ovina*)	13.0 (17.9)	9.0 (9.0)	6.0 (14.3)	1.0 (1.5)	6.0 (10.5)	4.0 (5.1)	14.0 (11.9)	<1.0 (0.4)		
black huckleberry (*Gaylussacia baccata*)	15.0 (32.7)	23.0 (30.0)	16.0 (28.1)	14.0 (19.1)	17.0 (22.1)	9.0 (12.0)	18.0 (27.5)	13.0 (13.2)		
bayberry (*Myrica pensylvanica*)	2.0 (6.0)	7.0 (12.9)	9.0 (17.8)	2.0 (34.6)	5.0 (17.4)	<1.0	4.0 (12.9)	4.0 (11.5)		
Virginia rose (*Rosa virginiana*)	4.0 (5.9)	4.0 (6.6)	3.0 (6.6)	<1.0 (2.3)	5.0 (10.4)	1.0 (2.8)	2.0 (4.7)	<1.0 (1.1)		
swamp dewberry (*Rubus hispidus*)	12.0 (12.2)	18.0 (22.6)	22.0 (27.2)	3.0 (17.1)	12.0 (23.9)	5.0 (7.7)	8.0 (17.5)	2.0 (4.2)		
little bluestem (*Schizachyrium scoparium*)	40.0 (25.6)	11.0 (8.9)	37.0 (19.8)	26.0 (15.5)	43.0 (25.2)	26.0 (13.8)	49.0 (24.5)	28.0 (17.5)		
poison ivy (*Toxicodendron radicans*)	0.0	<1.0	<1.0	0.0	4.0 (13.2)	1.0 (4.8)	0.0	0.0		
lowbush blueberry (*Vaccinium angustifolium*)	5.0 (17.5)	11.0 (33.2)	4.0 (9.1)	8.0 (30.9)	12.0 (21.5)	19.0 (28.0)	15.0 (31.7)	16.0 (29.4)		
Total No. of Species in Plot	25	29	34	41	24	34	30	34		
Average No. of Spp./Quadrat	5.8 (1.9)	6.6 (2.1)	7.6 (2.2)	10.2 (2.8)	6.8 (1.6)	9.2 (2.0)	6.3 (2.3)	8.0 (3.1)		

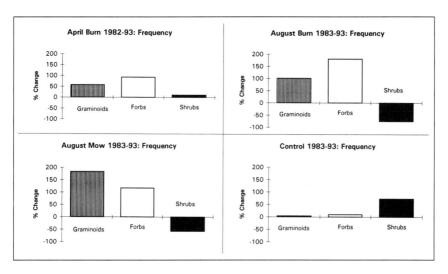

Figure 1. Percent change in graminoid (gray), forb (clear), and shrub (black) cover in treatment and control plots between 1982 and 1993 in Ram Pasture, Nantucket I., MA.

move along different trajectories. This analysis depicts the control plot moving in an opposite direction to that of the three treatment plots, corresponding to the opposite trends in the frequencies of woody and herbaceous species.

Overall species diversity, as measured by the number of species per line-segment, increased between 26 and 36 percent in the three treatment plots over pretreatment values, compared to a 15 percent increase in the control (Table 1). Forbs and graminoids accounted for most of these increases.

Although the dominant taxa in all of the plots remained the same, individual species exhibited widely varied responses to the type and season of treatments, even within groups with generally similar growth forms (Tables 1 and 2). Annual data from several species illustrate this. Among the graminoids, for example, little bluestem and Pennsylvania sedge (*Carex pensylvanica*) displayed different trends. Little bluestem declined in cover but remained largely unchanged in frequency in all of the plots (Table 1). However, the pattern of the cover decline differed among treatments (Fig. 4). In the control plot, the decline was steady throughout the period. A similar but less rapid decline occurred in the August mow plot. The two burn plots showed a more stepped response; after an initial decline in little bluestem cover following the first burn in the August burn plot, or following the first two burns in the April burn plot, values remained relatively constant.

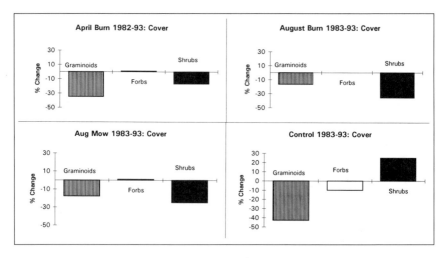

Figure 2. Percent change in graminoid (gray), forb (clear), and shrub (black) frequency in treatment and control plots between 1982 and 1993 in Ram Pasture, Nantucket I., MA.

In contrast to little bluestem, Pennsylvania sedge cover remained relatively unchanged in the treatment plots and declined only in the control plot (Table 1). However, this species spread widely in all of the treatment plots, as indicated by the gradual increase in frequency values (Fig. 4, Table 2). As in little bluestem, the height of Pennsylvania sedge also declined 20 percent in the August burn plot but remained unchanged in the April burn plot.

Widely different responses between species to treatments were also observed in height data (Fig. 5). For example, the height of little bluestem exhibited a high annual variability in the August treatments, increasing an average of 24 percent (August mow) to 38 percent (August burn) in the year following each burn. No such response was observed in the April burn or control plots. Pennsylvania sedge (not depicted) showed an opposite pattern, declining about 20 percent in height in the year following each April or August burn. The top-killing and subsequent resprouting of shrubs following each treatment is illustrated by black huckleberry (Fig. 5). Declines in the height of this species in all three treatment plots were similar, and contrasted markedly with the gradual height increase in the control plot.

Three different responses of morphologically similar clonal shrubs are illustrated in data from the August burn plot (Fig. 6). Lowbush blueberry (*Vaccinium angustifolium*) gradually increased with repeated burning, closely matching its behavior in the control plot. Black huckleberry showed little change over the 11 years but declined in cover immediately following each burn. Bayberry exhibited a third

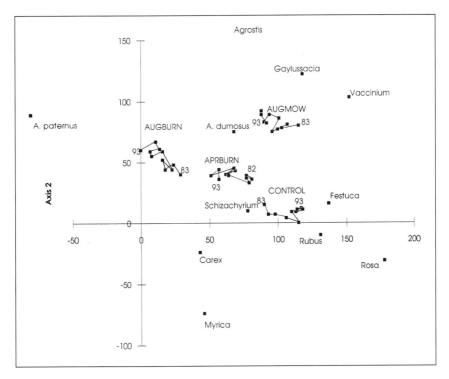

Figure 3. Detrended correspondence analysis of frequency data from all plots in Ram Pasture, Nantucket I., MA. Locations of species that received the greatest weights in this analysis are labeled by genus (except bushy aster [Aster dumosus]); treatments are capitalized.

pattern, declining and recovering following each of the first two burns, then falling to a low level following the third burn (1987).

The effects of these treatments on rare herbaceous taxa are of interest in developing methods for enhancing their populations. The frequency of one such species, sandplain blue-eyed grass, increased from 0 to 23 percent in the August burn plot over the course of this study. Other rare species were not encountered with sufficient frequency to ascertain whether statistically significant changes occurred. By resampling all 150 segments originally sampled in the April burn plot, however, we were able to evaluate effects on three species listed as being of special concern by the Massachusetts Natural Heritage and Endangered Species Program (1994). As in the August burn plot, sandplain blue-eyed grass also increased in the April burn plot, from 1 percent in 1982 to 6 percent in 1994. The frequency of sandplain flax increased from 1 to 12 percent, and bushy rockrose declined slightly, from 14 to 11 percent. Several hundred individuals of one state-listed

Table 2. Percent frequency of the major species in plots in Ram Pasture, Nantucket I., MA. The 1982 and 1983 data are pretreatment. All plots were treated in odd years, beginning in 1983. The April burn plot received six treatments, whereas the August plots received five.

Species	Control		August Burn			August Mow		April Burn	
	1983	1993	1983	1993	1983	1993	1982	1993
agrostis (*Agrostis* spp.)	30	13	20	27	34	45	37	7
bearberry (*Arctostaphylos uva-ursi*)	0	0	27	27	0	3	3	7
purple chokeberry (*Photinia floribunda* [= *Aronia arbutifolia*])	7	7	7	7	24	24	17	23
bushy aster (*Aster dumosus*)	7	60	50	70	59	86	40	70
small flat-topped aster (*Aster paternus*)	7	7	43	77	10	28	17	37
sedge (*Carex pensylvanica/umbellata*)	57	73	30	87	24	72	23	77
poverty grass (*Danthonia spicata*)	17	7	23	23	0	48	7	7
goldenrod (*Euthamia tenuifolia/graminifolia*)	0	7	7	0	21	28	0	3
sheep fescue (*Festuca ovina*)	63	90	50	63	55	97	77	77
black huckleberry (*Gaylussacia baccata*)	30	37	37	37	62	59	43	47
bayberry (*Myrica pensylvanica*)	27	33	50	7	28	3	33	30
Virginia rose (*Rosa virginiana*)	43	50	37	33	59	28	40	27
swamp dewberry (*Rubus hispidus*)	57	77	60	37	66	69	47	43
sheep sorrel (*Rumex acetosella*)	3	0	17	23	14	34	3	3
little bluestem (*Schizachyrium scoparium*)	97	87	87	97	100	100	90	97
poison ivy (*Toxicodendron radicans*)	0	7	3	0	28	21	0	0
lowbush blueberry (*Vaccinium angustifolium*)	17	33	17	20	31	45	47	60

endangered species that had been seen in only one other location on Nantucket, papillose nutsedge (*Scleria pauciflora* var. *caroliniana*), appeared in the August burn plot after three treatments (1988).

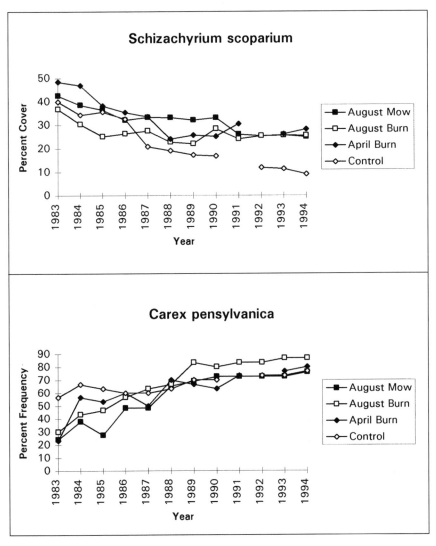

*Figure 4. Percent cover of little bluestem (*Schizachyrium scoparium*) and percent frequency of Pennsylvania sedge (*Carex pensylvanica*) in each of the four plots in Ram Pasture, Nantucket I., MA. Pretreatment values are depicted in 1983, with treatments occurring in all odd-numbered years.*

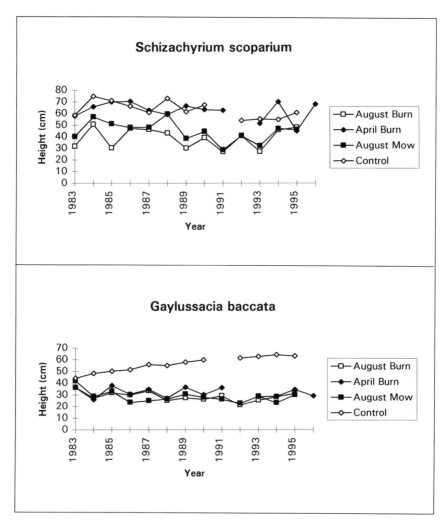

Figure 5. Height of little bluestem (Schizachyrium scoparium) *and black huckleberry* (Gaylussacia baccata) *in each of the four plots in Ram Pasture, Nantucket I., MA. Pretreatment values are depicted in 1983, with treatments occurring in all odd-numbered years.*

Discussion

The increases in species diversity in all three treatment plots resulted from higher frequencies of herbaceous species, primarily forbs in the burn plots and graminoids in the mow plot (Fig. 2). All plots were dominated by native species. Only two nonindigenous species occurred with any frequency—sheep fescue and sheep sorrel (*Rumex acetosella*)—both of which increased most markedly in the August

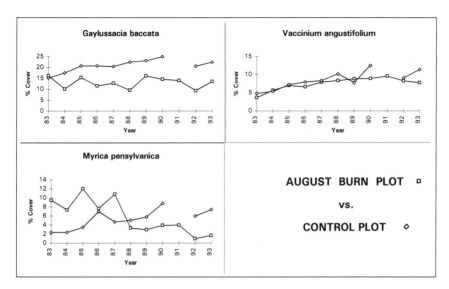

*Figure 6. Percent cover of black huckleberry (*Gaylussacia baccata*), lowbush blueberry (*Vaccinium angustifolium*), and bayberry (*Myrica pensylvanica*) in the August burn and control plots 1983-1993 in Ram Pasture, Nantucket I., MA (1991 data are missing in control plot because of a hurricane).*

mow plot. The proliferation of herbaceous taxa may have been caused by several factors. Reduction in the height, cover, and vigor of many of the taller shrubs in all treatments probably reduced competition for light, and perhaps other resources; accumulations of thatch, litter, and the presence in some areas of a lichen cover appeared to inhibit germination and growth of many herbs; burning reduced these components, allowing species to germinate or expand into the formerly covered areas; and finally, flowering and fruiting of many herbaceous species, stimulated by burning, may have provided an increased seed source to germinate in favorable sites. Reduced competition from shrubs was probably the most important factor, since the greatest negative impacts on shrubs coincided with the greatest increases in herbaceous taxa.

These results suggest that burning and mowing are viable tools for maintaining sandplain grasslands. Herbaceous species richness is increased, numbers of rare species are maintained or enhanced, and many shrubs are reduced in cover and frequency, especially by summer treatments. Despite the fact that summer burns are generally of low intensity, they appear to be more effective in reducing shrubs than spring burns, which tend to burn completely and with greater intensity. It is noteworthy, however, that even after six treatments in 11 years, shrubs continued to persist on these plots. This suggests that when restoration objectives dictate the widespread removal of shrubs, more aggressive measures than those employed in this study may be necessary, such as applying herbicides or combining burning and mowing.

Caution should always be exercised in extrapolating results of studies from one area to another. However, the work of Niering and Dreyer (1989) provides several contrasts with our results. They examined the effects of annual and biennial spring burns on Connecticut grasslands dominated by little bluestem. After 12 annual spring burns, they noted that clonal shrubs such as black huckleberry increased in cover and frequency at rates equal to or exceeding those found in the reference plots. This difference from our results may be due to more complete or hotter burns applied in our study, given two years to accumulate fuels between fires, or it may reflect the greater proximity of the Ram Pasture sites to the ocean, where salt spray may slow the rate of shrub regrowth. The sharp decline in bayberry in our August burn plot in 1992 (Fig. 6) may reflect the combined impacts of burning and salt spray from a hurricane that struck Nantucket in August 1991.

The productivity of little bluestem has been reported by Niering and Dreyer (1989) to increase in sites burned annually in Connecticut, as well as by others (Jordan 1965, Swan 1970) following burns in the eastern United States. Although no direct measurements of productivity were made in our study, neither cover nor height (both of which would be expected to be positively correlated with productivity) of little bluestem increased in either of the burn plots. Contrasting effects of annual versus biennial burning may explain these differences. Tester (1989), Gimingham (1987), and others have found that less frequent burns tend to be of greater intensity and may have a negative impact on some species that normally benefit from frequent burning.

Conclusions

Of the three treatments we tested, summer burning was the preferred method for managing Nantucket's native grasslands to encourage a diversity of herbaceous species and to limit encroachment of several shrub species. Summer mowing yielded similar results, although grasses may have been favored more than forbs. Spring burning was less effective but may be an important alternative when summer burning is not an option.

Acknowledgments

We thank the Nantucket Conservation Foundation for making the site available for this study and for providing logistical support in mowing the test plots and fire breaks. The Nantucket Fire Department provided assistance in carrying out the prescribed burns. We are grateful to all the volunteers who assisted in data collection and who served on burn crews.

Literature Cited

Dunwiddie, P. W. 1991. Comparisons of above-ground arthropods in burned, mowed, and untreated sites in sandplain grasslands. Am. Midl. Nat. 125: 206-212.

Dunwiddie, P. W., and C. Caljouw. 1990. Prescribed burning and mowing of coastal heathlands and grasslands in Massachusetts. N. Y. State Mus. Bull. 471: 271-275.

Dunwiddie, P. W., W. A. Patterson III, and R. E. Zaremba. 1995. Evaluating changes in vegetation from permanent plots: an example from sandplain grasslands in Massachusetts. Pp. 245-250 *in* Ecosystem monitoring and protected areas (T. B. Herman, S. Bondrup-Nielsen, J. H. Martin Willison, and N. W. P. Munro, eds.). Sci. and Management of Protected Areas Assoc., Wolfville, NS.

Dunwiddie, P. W., R. E. Zaremba, and K. A. Harper. 1993. Classification of coastal sandplain grasslands and heathlands in Massachusetts. Report to Massachusetts Natural Heritage and Endangered Species Prog., Boston.

Gauch, H. G. 1982. Multivariate analysis in community ecology. Cambridge Univ. Press, Cambridge, U.K.

Gimingham, C. H. 1987. Harnessing the winds of change: heathland ecology in retrospect and prospect. J. Ecology 75: 895-914.

Jordan, C. 1965. Fire influence in old fields of the New Jersey Piedmont. N. J. Acad. Sci. Bull. 10: 7-12.

Massachusetts Natural Heritage and Endangered Species Program. 1994. Massachusetts list of endangered, threatened and special concern species. Massachusetts Natural Heritage and Endangered Species Prog., Boston.

Niering, W. A., and G. D. Dreyer. 1989. Effects of prescribed burning on *Andropogon scoparius* in postagricultural grasslands in Connecticut. Am. Midl. Nat. 122: 88-102.

Swan, F. R. 1970. Post-fire response of four plant communities in south-central New York state. Ecology 51: 1074-1082.

Tester, J. R. 1989. Effects of fire frequency on oak savanna in east-central Minnesota. Bull. Torr. Bot. Club 116: 134-144.

Peter W. Dunwiddie: Center for Biological Conservation, Massachusetts Audubon Society, Lincoln, MA 01773 (present address: The Nature Conservancy, Washington Field Office, Seattle, WA 98101).

William A. Patterson III: Department of Forestry and Wildlife Conservation, University of Massachusetts, Amherst, MA 01003.

James L. Rudnicky: Department of Forestry and Wildlife Conservation, University of Massachusetts, Amherst, MA 01003.

Robert E. Zaremba: The Nature Conservancy, Albany, NY 12204.

Experimental Use of Prescribed Fire for Managing Grassland Bird Habitat at Floyd Bennett Field, Brooklyn, New York

James L. Rudnicky, William A. Patterson III, and Robert P. Cook

Abstract

Floyd Bennett Field (FBF), an operational airport in Brooklyn, NY, until the early 1970s, is now managed as a national recreation area. Fields between the runways consist of little bluestem (*Schizachyrium scoparium*) grasslands and are important breeding grounds for regionally rare grassland birds. In the mid-1980s, concern over loss of habitat because of invasion by woody and exotic plant species led to management action. From 1985 to 1990, 55 hectares (ha) were cleared of woody vegetation. Since 1990, this grassland has been mowed annually.

In 1993 prescribed fire was tested as an alternative habitat-management technique. Two fields were burned: one in April and the other in August. Areas adjacent to each of these two fields were mowed in late August. Vegetation in both the burned and mowed areas was surveyed in March and July; the field burned in August was also surveyed in November.

In July 1993 on the April burn field there were significant increases in litter cover and bare soil in the burned area compared with the adjacent mowed area. Furthermore, little bluestem height increased significantly on the April burn site. By November, the field burned in August showed significantly greater bare soil, dewberry (*Rubus flagellaris*), and little bluestem cover than the adjacent mowed area. Dewberry was significantly taller in the burned area, whereas little bluestem was significantly taller in the mowed area. The following summer, in June 1994, the August burn site still had a high percentage of bare soil cover, whereas in the April burn site, bare soil had diminished and was similar to the adjacent mowed area. There was less regrowth of woody plants in the August burn site than in the April burn site.

Burning, especially in August, increased bare soil cover, which should improve habitat for grassland breeding birds. Future sampling of the vegetation and breeding-bird populations will be needed to determine how long these habitat improvements persist.

99

Introduction

Floyd Bennett Field (FBF), a unit of Gateway National Recreation Area in Brooklyn, NY, is a former airport with grassy fields between the paved runways. These fields are used as breeding sites by regionally rare grassland birds, including Grasshopper Sparrows (*Ammodramus savannarum*; Lent and Litwin 1989), for whom grasses and herbaceous vegetation, plus open bare ground, are important components of breeding habitat (Whitmore 1981). To improve nesting habitat, the fields at FBF have been mowed annually since 1985. Mowing, however, has not eliminated woody plant species and has created a thick organic thatch on top of the soil. In West Virginia, Whitmore (1979) found that a thick litter layer and lack of bare ground were correlated with a population decline in breeding Grasshopper Sparrows. In an effort to create better breeding conditions, parts of the fields at FBF were burned in 1993 to determine if fire could be used safely and effectively to manage the habitat.

Two fields were selected for burning (Fig. 1), one in April and one in August. We wanted to determine if burning at different times of the year would affect the vegetation differently. Spring burns are generally easier to ignite because fine fuel moistures are low; also, a large portion of a plant's resources are below ground then and can be used to build new tissues following the fire (Salisbury and Ross 1985). Summer burns are harder to ignite because live fuels have high moisture contents; and root resources may be at their lowest levels in summer, which can hamper vegetative regrowth. Because different plant species have different growth patterns, species may be affected differently depending on when fires are applied.

To document changes in vegetation structure at FBF, we conducted surveys before and after burning in the two burned fields as well as in adjacent unburned control areas. This paper compares the data to determine how burning affected the vegetation communities and whether spring or summer burning was a more effective management technique.

Methods

Data Collection

The runways at FBF divide the fields into separate management sites. Sites C (10 ha) and G (16.5 ha) were selected for treatment, and approximately half of each site was burned. Site C was burned on 8 April 1993 and site G on 4 August 1993. All of site C and the unburned half of site G were mowed in late August 1993.

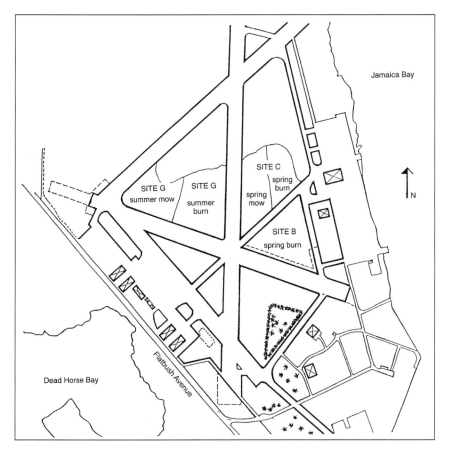

Figure 1. Map of Floyd Bennett Field, Brooklyn, NY, showing site locations (1 cm = 115.4 m).

Data were collected along transects using two sampling techniques. The first involved point-intercept sampling (Mueller-Dombois and Ellenberg 1974) at 1-meter (m) intervals and measuring the height (in centimeters [cm]) of the tallest plant of each species. Bare soil, litter, or any other nonplant feature was recorded (and given a height of 1) if no vegetation covered the point. Unless stated otherwise, these features are referred to as "species" when mentioned in the remainder of this paper; mosses and lichens are also each referred to as "species." A preliminary sample of the sites in 1992 identified three vegetation types. During the March 1993 survey, each sampling point was assigned to one of these types: "little bluestem," "dewberry," or "other grasses." The number of sampling points, by site, within each type is given in Table 1.

J. L. Rudnicky et al.

Table 1. *Distribution of sampling points by site, treatment, and vegetation type at Floyd Bennett Field, Brooklyn, NY, 1993. Some points in site C were lost because of construction of the fireline.*

Treatment	Vegetation Type	Site C (April)	G (August)
Mow	little bluestem	445	191
	dewberry	109	294
	other grasses	20	91
	(total number of points)	574	576
Burn	little bluestem	538	229
	dewberry	72	246
	other grasses	108	101
	(total number of points)	718	576
	(total number of points)	1292	1152

The second sampling method involved collecting cover data within a square plot frame (0.16 m^2). Because more transects were established in site C, the plots were located at 25-m intervals, whereas they were at 20-m intervals in site G. This ensured that the same number of plots was sampled between sites. Within each plot, cover was visually estimated to the nearest 10 percent for each species. Species with a cover of less than 5 percent were assigned a value of 1 percent. Vegetation type was also recorded for each plot.

Transects 96 m long were permanently located in both sites in March 1993 to sample the vegetation and nonplant features. In site C 16 transects were established, 8 in the burned area and 8 in the mowed area. In site G, 12 transects were established, 6 in the burned area and 6 in the mowed area. All transects were surveyed in March and July. A partial survey was conducted in site G in November; all of the burned transects but only half of the mowed transects were sampled.

Data Analyses

Data were analyzed separately for each site because of differences in treatment history and species composition. Species cover, calculated as the number of points on which a species occurred divided by the total number of points sampled, was summarized by vegetation types for each month (March, July, and November). Thus, a species found on 50 out of 100 points had a cover of 50 percent. Importance values (IVs) were also calculated for the point intercept data. IVs are the sum of relative cover (a species' mean cover divided by the sum of the mean cover of all species) and relative height (mean maximum height divided by sum of the means of all heights). A maximum IV value is 200. IVs were calculated separately for the March, July, and November data for each species found in each vegetation type.

Analysis of variance (ANOVA) tests were conducted on species cover, IV, and height data, testing for differences between treatments, vegetation type, and treatment-vegetation type interaction (Dixon et al. 1988). The data were normalized before the ANOVAs were conducted. If a significant difference ($p < 0.05$) was found, a t-test with a Bonferroni adjusted p-value (Dixon et al. 1988) was used to test for differences between vegetation types across treatments. Analyses were conducted separately for the March, July, and November samples.

Species data collected on the 0.16-m^2 plots were used in a detrended correspondence analysis (DCA; Gauch 1982) using the computer program CANOCO (Ter Braak 1988). These results were used to identify trends in species cover among the plots. Cover values estimated in the field for each plot (as opposed to cover values calculated from point intercept) were used, with rare species downweighed. Analyses were conducted separately for data from sites C and G. In these analyses, data from March, July, and November were combined into one data set for each site. A correlation analysis using Pearson's correlation coefficients was conducted between sample DCA scores for each axis and species cover data to identify possible species cover patterns along the two axes.

1994 Subsample

In June 1994 a subset of 96 plots was sampled along transects in the burned and mowed areas of both sites. This was done to provide managers with an indication of how the sites had regrown the year after treatment. Mean values of percent cover were calculated for selected species. Comparative analyses were not conducted because of the small sample size; these results are included, however, because the data show several interesting patterns. The sample size necessitated that the data be summarized by treatment and not vegetation type.

Results

Changes in Species Importance in 1993

While sampling vegetation, we identified 71 species (Table 2). Within any vegetation type, fewer than four species were dominant (Table 3). In March 1993, little bluestem dominated the little bluestem types in the mowed and burned areas (sites C and G). Dewberry was the dominant species in the dewberry types, with litter and little bluestem also being important features of this type. Sites C and G differed in the composition of the other grasses type, a general category composed of species unique to each site (e.g., love grass [*Eragrostis curvula*] and Japanese knotweed [*Polygonum cuspidatum*] in site C and common mugwort [*Artemisia vulgaris*] in site G).

By July, more species had appeared which resulted in the IVs being more similar. The mowed little bluestem types were still dominated by little bluestem, but dewberry and common blackberry (*Rubus allegheniensis*) IVs had increased. Litter IVs did not change appreciably from March to July. In the dewberry type, litter importance decreased because of increases in plant foliage. In the other grasses type, red fescue (*Festuca rubra*) became the dominant species. In the burned half of site C, the little bluestem type showed an increase in IV for bare soil when compared to March (preburn). Bare soil had low IVs in July in the other sites. In areas of burned dewberry in site C, dewberry and common blackberry IVs decreased from March to July but were similar to the mowed dewberry IVs in site C.

In November only site G was surveyed. Half of this site was burned in early August, whereas the other half was mowed, as part of the regular FBF management schedule, in late August. In the mowed half, November species IVs were similar to March IVs in the three vegetation types. Large changes in IVs were found in all types in the burned area. In each vegetation type, bare soil IVs increased as litter IVs decreased. Dewberry increased in importance in the little bluestem and other grasses types but remained relatively unchanged in the dewberry type. Little bluestem IVs were reduced from March and July values in all types.

Analysis of Variance

The full analysis table is included in the Appendix. In March, most differences in species values were between vegetation types and not between treatments and control sites. This was to be expected because the types were defined by species cover values. In both sites, the greatest cover and IV were from litter in the mowed and burned

Burning Grassland Bird Habitat in New York

Table 2. *Species list for 1993 field sampling at Floyd Bennett Field, Brooklyn, NY (nomenclature from Fernald 1950). Nonplant features encountered in the field are also included.*

yarrow (*Achillea millefolium*)	litter
quack grass (*Agropyron repens*)	mosses
tickle grass (*Agrostis hyemalis*)	bayberry (*Myrica pensylvanica*)
agrostis (*Agrostis* sp.)	yellow wood sorrel (*Oxalis stricta*)
field garlic (*Allium vineale*)	panic grass (*Panicum* spp.)
common ragweed (*Ambrosia artemisiifolia*)	woodbine (*Parthenocissus quinquefolia*)
pearly everlasting (*Anaphalis margaritacea*)	reed (*Phragmites communis*)
common mugwort (*Artemisia vulgaris*)	Japanese knotweed (*Polygonum cuspidatum*)
common milkweed (*Asclepias syriaca*)	lady's thumb (*Polygonum persicaria*)
many-flowered aster (*Aster ericoides*)	rough-fruited cinquefoil (*Potentilla recta*)
aster (*Aster* sp.)	black cherry (*Prunus serotina*)
oat (*Avena* sp.)	dwarf sumac (*Rhus copallina*)
gray birch (*Betula populifolia*)	poison ivy (*Rhus radicans*)
sedge (*Carex* sp.)	multiflora rose (*Rosa multiflora*)
climbing bittersweet (*Celastrus scandens*)	common blackberry (*Rubus allegheniensis*)
thistle (*Cirsium* sp.)	dewberry (*Rubus flagellaris*)
Queen Anne's lace (*Daucus carota*)	cut-leaved blackberry (*Rubus lacinatus*)
love grass (*Eragrostis curvula*)	field sorrel (*Rumex acetosella*)
purple love grass (*Eragrostis spectabilis*)	common elderberry (*Sambucus canadensis*)
daisy fleabane (*Erigeron annuus*)	bouncing bet (*Saponaria officinalis*)
hair-like fescue (*Festuca capillata*)	little bluestem (*Schizachyrium scoparium*)
meadow fescue (*Festuca elatior*)	soil
giant fescue (*Festuca gigantea*)	bittersweet nightshade (*Solanum dulcamara*)
fescue (*Festuca myuros*)	Canada goldenrod (*Solidago canadensis*)
sheep fescue (*Festuca ovina*)	lance-leaved goldenrod (*Solidago graminifolia*)
red fescue (*Festuca rubra*)	early goldenrod (*Solidago juncea*)
sweet everlasting (*Gnaphalium obtusifolium*)	rough-stemmed goldenrod (*Solidago rugosa*)
grass (gramanoid sp.)	seaside goldenrod (*Solidago sempervirens*)
soft rush (*Juncus effusus*)	goldenrod (*Solidago* sp.)
red cedar (*Juniperus virginiana*)	showy goldenrod (*Solidago speciosa*)
wild lettuce (*Lactuca canadensis*)	slender-leaved goldenrod (*Euthamia tenuifolia*)
prickly lettuce (*Lactuca scariola*)	common dandelion (*Taraxacum officinale*)
bush clover (*Lespedeza cuneata*)	poison ivy (*Toxicodendron radicans*)
lichens	tree stump
blue toadflax (*Linaria canadensis*)	common mullein (*Verbascum thapsus*)
butter-and-eggs (*Linaria vulgaris*)	

Table 3. Species Importance Values (IVs) by site, vegetation type, and month for Floyd Bennett Field, Brooklyn, NY, 1993. Site C was not sampled in November. Maximum value (= sum of all IVs) is 200.

	Site C (April)			Site G (August)		
	Mar	July		Mar	July	Nov
Mowed – Little Bluestem						
little bluestem	141	69	little bluestem	134	88	100
litter	27	28	litter	39	30	35
dewberry	17	19	grass	13	5	25
grass	5	9	dewberry	9	23	11
common blackberry	0	24	red fescue	0	13	2
			panic grass spp.	0	6	18
Mowed – Dewberry						
little bluestem	57	25	dewberry	82	88	59
dewberry	55	25	litter	59	21	65
litter	44	21	little bluestem	24	16	32
bayberry	13	9	grass	12	2	3
common blackberry	0	42	bayberry	10	12	10
Mowed – Other Grasses						
grass	98	3	grass	107	11	116
litter	24	10	little bluestem	40	16	30
dewberry	22	21	litter	23	24	19
little bluestem	16	4	dewberry	14	23	14
red fescue	0	87	red fescue	0	65	0
Japanese knotweed	0	33				
rough-stemmed goldenrod	0	19				
Burned – Little Bluestem						
little bluestem	133	76	little bluestem	138	75	41
litter	34	37	dewberry	29	55	43
dewberry	14	20	litter	21	28	7
grass	7	7	grass	6	1	6
soil	0	11	red fescue	0	6	13
			soil	1	1	42
			goldenrod sp.	1	2	14
Burned – Dewberry						
dewberry	68	48	dewberry	81	81	87
litter	57	32	litter	53	23	2
little bluestem	34	15	little bluestem	31	24	8
bayberry	14	11	bayberry	13	12	1
bush clover	0	23	soil	2	1	29
common blackberry	0	18	goldenrod sp.	3	0	14
			grass	11	2	13
Burned – Other Grasses						
grass	55	15	grass	117	1	49
love grass	49	55	dewberry	24	33	32
little bluestem	40	26	little bluestem	23	8	4
litter	37	32	litter	22	29	1
			red fescue	0	43	23
			common mugwort	0	30	25
			soil	3	0	24

dewberry types. In site C (Fig. 2), litter importance in the mowed dewberry type was different than in the other mowed types. In site G (Fig. 3), litter cover and IV were also greater in the dewberry type.

Treatment differences were found in July in site C (Fig. 2). Little bluestem, red fescue, and common blackberry were taller in the mowed vegetation types than in the adjacent burned types. There was more exposed soil in the burned types, especially in the little bluestem types. Treatment differences in IVs were found for litter, red fescue, and common blackberry. Litter had higher values in all burned types, whereas red fescue and common blackberry had greater IVs in the mowed types. Litter cover increased between March and July in the burned little bluestem type.

In site G (Fig. 3), mowing and burning occurred in August. Most treatment differences were found in the November samples. In the mowed vegetation types, little bluestem was taller and had greater cover and IVs than in the adjacent burned types. Soil and dewberry

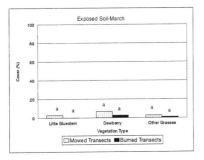
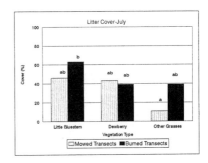
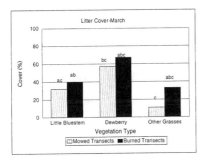
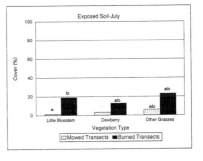

Figure 2. Exposed soil and litter cover in site C in March 1993 (preburn) and July 1993 (postburn). Within each chart, bars with different letters are significantly different at $p < 0.05$.

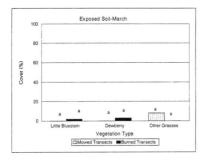
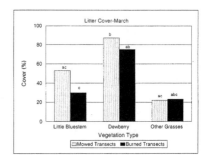
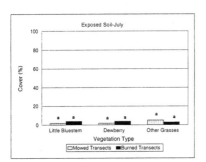
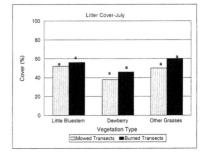
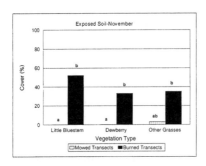
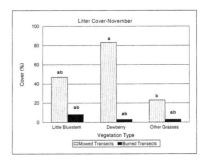

Figure 3. Exposed soil and litter cover in site G in March 1993 (preburn), July 1993 (preburn), and November 1993 (postburn). Within each chart, bars with different letters are significantly different at p < 0.05.

had greater cover and IVs in the burned types. Litter cover and IVs were larger in the mowed types.

Detrended Correspondence Analysis (DCA)

Plots of the first two DCA axes of site C sample scores (plus some of the species scores included to aid in interpretation) show that the first axis had high scores for daisy fleabane (*Erigeron annuus*), little bluestem, mosses, and common milkweed (*Asclepias syriaca*; Fig. 4). The lowest values were for seaside goldenrod (*Solidago sempervirens*), soft rush (*Juncus effusus*), common mugwort, and gray birch (*Betula populifolia*). Correlation analyses showed little bluestem to have the highest significant value (corr. = 0.69, $p < 0.001$) with the first axis. No strong negative correlations were found. Axis two had high positive scores for seaside goldenrod, red fescue, common mugwort, and Japanese knotweed. The lowest scores were for a sedge (*Carex* sp.) and for black cherry (*Prunus serotina*). Litter had the highest positive value (corr. = 0.31, $p < 0.002$), and the sedge species (corr. = -0.34, $p < 0.0001$) and sheep fescue (*Festuca ovina*; corr. = -0.32, $p < 0.001$) had the lowest values.

The first axis represents an increase in cover of little bluestem. Species with the lowest values on the first axis were found mostly in distinct patches that were not widely distributed throughout site C. The second axis shows a clear pattern in plot distribution as a function of species values. The highest scores were for species that generally form dense patches and exclude most other species. Sparsely distributed species had lower scores. High and low species scores along axis two typically occurred in the other grasses vegetation type. Thus, the second axis is mostly an expression of variability within this type. The weak positive correlation with litter may indicate increasing litter cover within samples along axis two.

In the mowed area, July samples were widely separated from March samples (Fig. 4). This was because new species were being identified in July. However, the plots from the burned area were not widely separated. When samples in the little bluestem and dewberry types were graphed separately (Fig. 4), the separation of July samples from March samples occurred mostly in the mowed little bluestem samples. This was because Japanese knotweed, red fescue, dewberry, common blackberry, and rough-stemmed goldenrod (*Solidago rugosa*) had greater cover on the mowed samples. In the burned area, the July samples had fewer new species growing on them; as a result, axes scores were closer to March samples. The July dewberry samples (Fig. 4) in the mowed and burned areas were separated from March samples in a similar manner.

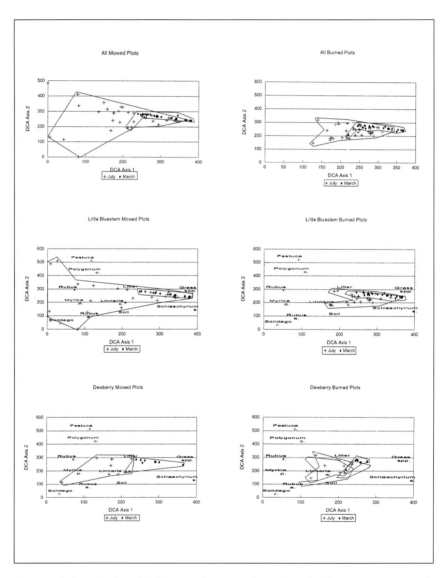

Figure 4. Site C: plot of DCA sample scores from March 1993 (preburn) and July 1993 (postburn) samples for all plots and for little bluestem and dewberry samples. Lines surrounding the plots were drawn arbitrarily to distinguish the spread of each month's samples.

Similar plots were made for the DCA based on site G cover data (Fig. 5). The highest first axis species scores were for a grass (*Agrostis* sp.), mosses, and little bluestem. The lowest scores were for poison ivy (*Toxicodendron radicans*), wild lettuce (*Lactuca canadensis*), and Japanese knotweed. The highest correlations were for little bluestem (corr. = 0.65, $p < 0.001$) and sweet everlasting (*Gnaphalium obtusifolium*; corr. = 0.60, $p < 0.001$). The lowest correlations were for dewberry (corr. = -0.53, $p < 0.001$) and Japanese knotweed (corr. = -0.53, $p < 0.001$). For the second axis, the highest values were for unidentified grasses, yellow wood sorrel (*Oxalis stricta*), a grass (*Agrostis* sp.), and a goldenrod (*Solidago* sp.). The lowest scores were for blue toadflax (*Linaria canadensis*), showy goldenrod (*Solidago speciosa*), and gray birch. The highest correlation was for a goldenrod species (corr. = 0.40, $p < 0.001$) and the lowest for little bluestem (corr. = -0.63, $p < 0.001$).

The first axis represents conditions indicating increasing cover of dewberry at the bottom of the axis and of little bluestem at the top of the axis. The second axis reflects decreasing little bluestem cover with increasing axis scores. For the mowed area, the location of July samples was widely separated from the March samples, but the November samples were once again close to the March samples. In contrast, by July the location of samples burned in April were separated from the March samples, but this separation persisted to November. For samples from the mowed dewberry and little bluestem types, the July dispersion was due to an increase in cover of dewberry, black cherry, bayberry (*Myrica pensylvanica*), and yarrow (*Achillea millefolium*). For samples from the burned areas, the dispersion in November reflected the increase in bare soil cover and in dewberry recovering more rapidly than other species. On the graphs of the burned areas, the November dewberry samples moved closer to the location of soil, whereas the little bluestem samples moved closer to the location of dewberry.

Results from 1994 Subsample

In 1994 (see Table 4) the amount of exposed soil remained nearly the same in the mowed areas but declined in the April burn site and increased in the August burn site. Litter cover increased in the mowed area and in the April burn site but decreased in the August burn site. Little bluestem cover decreased in all treatments, with the greatest decline occurring in the August burn site. Cover from unidentified grasses increased in all treatments. *Rubus* species cover increased in the mowed and April burn areas but decreased in the August burned site. Bayberry cover was less than 7 percent in 1993 and remained

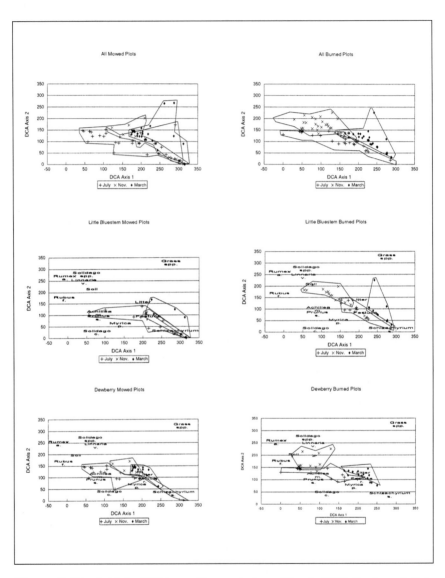

Figure 5. Site G: plot of DCA sample scores from March 1993 (preburn), July 1993 (preburn), and November 1993 (postburn) samples for all plots and for little bluestem and dewberry samples. Lines surrounding the plots were drawn arbitrarily to distinguish the spread of each month's samples.

nearly the same in the mowed and April burn areas; the species was not present in the August burn site.

Discussion

The answer to the question of whether mowing or burning best maintains grasslands depends on the desired community features. At FBF, the grassy areas between the runways provide breeding habitat for grassland-dependent birds, which generally require grassy clumps interspersed with bare ground (Whitmore 1981). Without management, plant succession at FBF would eventually result in these fields being dominated by shrub patches or early-successional forests. Mowing might prevent the grassy fields from being covered by woody plants, but other features that breeding birds require would be lost. For example, bare ground would be reduced, undesirable nonwoody species might become established, and preferred plant species would be reduced (Whitmore 1979). To prevent losing desirable habitat features, monitoring programs should be instituted to provide managers with information on how various management activities (including mowing and burning) alter the habitat.

The experiments conducted at FBF demonstrated that season of burn can yield different results. The DCA for July showed that the postburn species composition in the burned half of site C (April burn) resembled preburn conditions. In contrast, the postburn species composition of the half of site G burned in August (August burn) was very different from preburn conditions. Interestingly, in July the mowed half of site C differed more from preburn conditions than the

Table 4. Mean percent cover for selected site characteristics and species for areas burned and/or mowed in 1993 at Floyd Bennett Field, Brooklyn, NY. Values are based on 96 points sampled in each management plot in July 1993 and June 1994. The spring burn was burned in April 1993 and mowed in August 1993. The summer burn was burned in August 1993 and was not mowed.

	Mow		Spring Burn		Summer Burn	
	1993	1994	1993	1994	1993	1994
bare soil	4.1	6.3	19.8	5.2	0.0	42.7
litter	62.5	90.6	62.5	85.4	63.5	43.8
little bluestem	67.8	49.0	46.9	43.8	58.3	13.5
unidentified grasses	8.3	15.6	20.8	22.9	7.3	20.8
Rubus spp.	19.8	38.5	27.1	32.3	59.4	26.0
bayberry	4.2	4.1	5.2	6.3	2.1	0.0

burned half, whereas in November the mowed half of site G was similar to preburn conditions.

Spring burning may have inhibited or prevented the growth of plant species that typically become apparent during the summer, whereas burning in August appeared to open new spaces for different species to colonize. These findings suggest that cover of little bluestem may be maintained, perhaps through the reduction of cover of other species, by spring burning, whereas summer burning may reduce the cover of little bluestem and be more favorable to dewberry. However, little bluestem height was lower in the burned half of site C than in the mowed half, and this could have made little bluestem less competitive for light than other plants. In site C, summer litter cover was nearly the same as in the spring, but bare soil increased in the burned half. The summer of 1993 was very dry, and some plants that had resprouted following the burn died back and deposited new litter. The exposed soil occurred in areas not revegetated from the burn. In site G, postburn regrowth was not extensive enough for new litter to cover the exposed soil.

The year following the burns, additional differences between the two burned sites became apparent. Because new litter had accumulated, the spring burn had less bare ground than the summer burn. Plant cover had returned to preburn levels in the spring burn, but this did not happen in the summer burn. In the spring burn, little bluestem cover had declined, whereas the cover of other grass species had increased. Shrub cover was greatly reduced in the summer burn.

Conclusions

One of the primary goals of habitat management at FBF is to provide habitat for nesting grassland birds. Mowing is partially achieving this goal, but accumulating thatch is unfavorable to some bird species and invasion of low shrubs and nonwoody exotics may degrade the habitat in the future. The application of experimental prescribed fires in 1993 suggested that spring burns may inhibit the growth of undesirable plant species during the summer. Litter accumulates following spring burning, but it is less dense than with mowing and may not pose a problem for foraging birds. The 1993 summer burn was unusual in that it was conducted under very dry conditions (approximately 1 cm of precipitation had fallen in the previous two weeks), which resulted in more plant material being burned than would have happened under wetter, perhaps more typical, conditions. In the following growing season, the spring burn increased the cover of litter and hardwood species, whereas the opposite occurred after the summer burn. Studies on grasslands on Nantucket Island, MA, have also shown that summer burning resulted in greater hardwood shrub and tree mortality than burning in other seasons (Dunwiddie et al. 1997). Thus, summer burns may be more beneficial to maintaining bird-nesting habitat at FBF than spring burns or mowing. Future vegetation monitoring will be needed to determine if the modifications produced by burning persist over time.

Acknowledgments

We thank the Yankee Candle Company for support in finishing this manuscript.

Literature Cited

Dixon, W. J., M. B. Brown, L. Engelman, M. A. Hill, and R. I. Jennrich. 1988. BMDP statistical software manual. Vol. 1. Univ. of California Press, Berkeley.

Dunwiddie, P. W., W. A. Patterson III, J. L. Rudnicky, and R. E. Zaremba. 1997. Vegetation management in coastal grasslands on Nantucket Island, Massachusetts: effects of burning and mowing from 1982 to 1993. Pp. 85-98 *in* Grasslands of northeastern North America: ecology and conservation of native and agricultural landscapes (P. D. Vickery and P. W. Dunwiddie, eds.). Massachusetts Audubon Soc., Lincoln.

Fernald, M. L. 1950. Gray's manual of botany. 8th ed. American Book Co., New York.

Gauch, H. G., Jr. 1982. Multivariate analysis in community ecology. Cambridge Univ. Press, Cambridge, U.K.

Lent, R. A., and T. S. Litwin. 1989. Bird-habitat relationships as a guide to ecologically-based management at Floyd Bennett Field, Gateway National Recreation Area. Part I, baseline study. Seatuck Research Prog., Cornell Univ. Lab. Ornithol., Islip, NY.

Mueller-Dombois, D., and H. Ellenberg. 1974. Aims and methods of vegetative ecology. John Wiley, New York.

Salisbury, F. B., and C. L. Ross. 1985. Plant physiology. 3d ed. Wadsworth Publ. Co., Belmont, CA.

Ter Braak, C. J. F. 1988. CANOCO: a FORTRAN program for canonical community ordination by partial detrended canonical correspondence analysis, principal components analysis and redundancy analysis. Version 2.1. Tech. report LWA-88-02, Agricultural Mathematics Group, Wageningen, The Netherlands.

Whitmore, R. C. 1979. Short-term change in vegetation and its effect on Grasshopper Sparrows in West Virginia. Auk 96: 621-625.

Whitmore, R. C. 1981. Structural characteristics of Grasshopper Sparrow habitat. J. Wildl. Manage. 45(3): 811-814.

James L. Rudnicky: Department of Forestry and Wildlife Conservation, University of Massachusetts, Amherst, MA 01003.

William A. Patterson III: Department of Forestry and Wildlife Conservation, University of Massachusetts, Amherst, MA 01003.

Robert P. Cook: National Park Service, Gateway National Recreation Area, Floyd Bennett Field, Brooklyn, NY 11234.

Appendix. ANOVA summary by month, site, treatment, and vegetation type testing for differences in species height (cm), percent cover, and Importance Value at Floyd Bennett Field, Brooklyn, NY, in 1993. Only species with significant differences are listed. Values shown are species means. The analyses were conducted by treatment (Trt), vegetation type (Veg), and treatment x vegetation type interaction (Int). Differences are shown for $p < 0.05$ (*), $p < 0.01$ (**), and $p < 0.001$ (***). NA represents places where an analysis could not be conducted. For each line of species, treatments and vegetation types (e.g., Mow Blue) with different letters are significantly different at $p < 0.05$.

Species	p-Values Trt	Veg	Int	Mow Blue	Mow Dew	Mow Grass	Burn Blue	Burn Dew	Burn Grass
March - Site C - Height									
RUBFLA	***			1.10abc	3.49bc	0.41ac	1.17ab	3.82abc	0.21c
SCHSCO	***		**	9.60a	3.79b	0.46b	10.97a	2.07b	2.60b
MYRPEN	**			0.12a	1.07ab	0	0.18ab	0.69b	0.27b
ERACUR	*			0	0.10a	0.56a	0.22a	0	2.38a
March - Site C - Cover									
Litter	***			32ac	57bc	11c	40ab	67abc	33abc
March - Site C - Importance Value									
Litter	***			27.1a	43.3b	9.1a	22.4ab	45.7ab	26.6ab
RUBFLA	***			18.6ab	54.1bd	5.3cd	16.4bc	66.0ab	4.4ad
SCHSCO	***		**	140.4a	52.8b	9.8b	138.7a	28.3b	46.2b
ERACUR	*			0	1.3a	8.2a	2.4a	0	2.7a
March - Site G - Height									
RUBFLA	***			0.62a	4.67b	0.72a	1.67a	3.68ab	1.17ab
SCHSCO	***			11.10a	1.43b	4.02ab	11.10a	2.07b	2.13b
March - Site G - Cover									
Litter	***			53ac	87b	22ac	30c	75ab	23abc
Soil		***		0a	0a	8a	2a	3a	0a
RUBFLA	***			5a	50b	10a	17ab	43b	12ab
SCHSCO	***			73a	13b	25b	80a	13b	18b
MYRPEN	**			0a	5a	0a	0a	3a	0a
ARTVUL	**			0a	0a	5a	0a	0a	3a
March - Site G - Importance Value									
Litter	***			39.8a	60.9bc	18.5ac	22.4a	62.0c	20.6ab
Soil	*			0.9a	1.0a	8.7a	0.5a	3.6a	0.6a
RUBFLA	***			9.4a	80.8b	11.8a	23.7ab	71.2b	18.4ab
SCHSCO	***			134.9a	24.2b	49.5b	139.8a	24.4b	30.5b
MYRPEN	**			0a	11.7a	0a	1.4a	10.5a	1.5a
ARTVUL	*			0a	0a	8.5a	0a	0a	3.9a
July - Site C - Height									
RUBFLA	*			4.04a	5.52a	2.89a	2.05a	4.14a	1.01a
SCHSCO	**	**		10.01a	6.10ab	2.78b	8.45a	1.44ab	4.61a
FESRUB	***	NA	NA	3.20abc	6.16ac	16.71a	0.46b	0	11.20ac
RUBALL	***	NA	NA	4.92a	8.79a	0a	0.56b	4.46ab	0a

Species	p-Values Trt Veg Int			Mow Blue	Mow Dew	Mow Grass	Burn Blue	Burn Dew	Burn Grass
July - Site C - Cover									
Litter		**		46ab	43ab	11a	63b	39ab	39ab
Soil	***			1a	3a	6ab	19b	13ab	23ab
RUBFLA		**		20a	30a	13a	16a	31a	23a
SCHSCO		**		52ac	29abc	11c	59a	11bc	33abc
PRUSER		**		0a	7a	0a	1a	4a	0a
FESRUB	***			8a	20a	48a	1a	0a	3a
July - Site C - Importance Value									
Litter	**	***		29.4a	22.4ab	4.7b	40.5a	25.9ab	25.1ab
Soil		*		1.5a	2.1ab	3.9b	11.7a	8.6ab	17.8ab
RUBFLA		***		24.2a	31.6a	11.8a	22.7a	47.9a	10.3a
SCHSCO		***		72.1ab	31.5a	9.8a	89.7b	16.6a	44.6ab
FESRUB	***			12.0ab	24.8a	58.8a	3.9b	0	3.9b
RUBALL	***			28.6a	41.3a	0	5.2b	33.2a	0
July - Site G - Height									
RUBFLA		***		4.23a	16.35b	4.65ab	9.00ab	13.34ab	7.00ab
SCHSCO		***	*	13.85a	3.10b	3.68ab	13.22a	3.54b	1.60ab
MYRPEN		**		0.18a	3.00a	0a	0.76a	2.07a	0.63a
FESRUB	***			3.43ab	0.42ab	18.00a	1.26b	0ab	11.20ab
July - Site G - Cover									
RUBFLA		***		23a	75a	20a	41a	66a	31a
SCHSCO		***		68a	15b	18b	71a	21b	10b
FESRUB		***		10ab	0a	47b	3a	0a	30a
July - Site G - Importance Value									
RUBFLA		***		31.4ac	97.3bc	20.4abc	51.0abc	92.3b	38.7c
SCHSCO		***		96.0a	18.7b	22.3b	86.4a	25.7b	11.5b
FESRUB	***			18.4ab	1.6ab	74.6a	6.3b	0a	45.2a
ARTVUL	**			0.6a	0a	28.9b	0a	0a	31.2a
November - Site G - Height									
RUBFLA		***		0.70a	2.67a	1.30a	1.52a	3.28b	1.63ab
SCHSCO	***	***		8.57ac	1.77abc	2.90abc	1.70c	0.33a	0.40bc
ARTVUL		**		0a	0a	0.43a	0a	0a	1.70a
November - Site G - Cover									
Litter	***	***	***	47ab	83a	23b	8ab	3ab	3ab
Soil	***			0a	0a	3ab	52b	33b	35b
RUBFLA	***	***		7a	33ab	10a	30a	57b	27b
SCHSCO	***	***		53ab	13ab	17ab	20a	3b	5ab
ARTVUL		*		0a	0a	10a	0a	0a	17a
November - Site G - Importance Value									
Litter	***	***	***	39.2a	71.4ab	18.0c	7.4c	3.7abc	2.4bc
Soil	***			1.7a	1.0ad	2.1ac	53.2bc	37.3abc	32.9cd
RUBFLA	**	*		12.2abc	65.8abc	14.5b	57.2a	108.2c	34.3a
SCHSCO	***	***		109.7a	34.9ab	32.5ab	50.4a	10.4b	7.7b
ARTVUL		**	*	0a	0a	8.5a	0a	0a	29.0a

Species Codes: RUBFLA: *Rubus flagellaris* (dewberry); SCHSCO: *Schizachyrium scoparium* (little bluestem); ARTVUL: *Artemisia vulgaris* (common mugwort); FESRUB: *Festuca rubra* (red fescue); MYRPEN: *Myrica pensylvanica* (bayberry); PRUSER: *Prunus serotina* (black cherry); RUBALL: *Rubus allegheniensis* (common blackberry); ERACUR: *Eragrostis curvula* (love grass).

History of Grasslands in the Northeastern United States: Implications for Bird Conservation

Robert A. Askins

Concern about the future of neotropical migratory birds has become a major conservation priority of government agencies and conservation organizations in New England. Most attention has focused on forest bird communities, which have a high diversity of neotropical migrants. Long-term studies have shown that numerous species of forest-dwelling migrants have experienced severe declines at widely separated sites in the northeastern United States since the 1950s (Terborgh 1989, Askins et al. 1990), so it is not surprising that research efforts and public interest have centered on these species. Despite the great alarm about the decline of forest migrants, however, the evidence for a broad, continentwide population decline is not convincing for most of these species. Although many of these species have declined in particular regions and during particular time periods, they have increased in other regions and during other time periods (James et al. 1992, Askins 1993, Peterjohn 1994).

In contrast to forest migrants, grassland bird species, some of which are neotropical migrants, generally have shown more consistent and severe population declines. Analysis of Breeding Bird Survey data indicates that 16 of the 19 species of grassland and savanna birds that breed in eastern North America have shown declining trends since 1966 and that 12 of these species have declined significantly (Askins 1993). Moreover, state lists of endangered and threatened bird species for the northeastern states include a large number of grassland specialists and very few forest specialists (Vickery 1992). Of the 40 species listed as endangered, threatened, or of special concern in three or more northeastern states, 13 are grassland or savanna specialists (using the habitat categories in Askins 1993), and only 3 (all raptors) are forest specialists. For example, Upland Sandpiper (*Bartramia longicauda*), Northern Harrier (*Circus cyaneus*), Loggerhead Shrike (*Lanius ludovicianus*), Grasshopper Sparrow (*Ammodramus savannarum*), Henslow's Sparrow (*A. henslowii*), and Vesper Sparrow (*Pooecetes gramineus*) are listed in all or most of the New England states (Vickery 1992). Many grassland species have declined throughout the eastern United States, both in the heavily forested Northeast and in the more agricultural Midwest (Herkert 1991, Bollinger and Gavin 1992).

Unlike forest migrants, grassland birds have received relatively little attention from most government agencies and conservation organizations in the Northeast. Despite well-documented population declines, there is surprising complacency, largely because of the general impression that most grassland species are not native to the region but instead invaded it from western savannas and prairies after the forests were cleared for agriculture. For example, Whitcomb (1987) argues that this invasion of the eastern "neosavanna" created by agriculture has been a "failed experiment for many of these species," which are now declining. The implication is that these declines are the result of a return to ecosystems more similar to those before European settlement and therefore should not be a cause for concern. According to Whitcomb, these species could survive only with active management to preserve grassland "in a region where [grassland] is inappropriate as an equilibrium community."

Another reason for the lack of concern about grassland species is reflected in a preliminary assessment of the status of particular migrant species in the northeastern United States (Smith et al. 1993). Although early-successional species, including grassland species, were disproportionately represented in the list of species with declining populations in this region, most grassland birds had relatively low ranks in terms of the overall threat of extinction. This was largely due to the perception that these species have high global abundance and that the Northeast is not important to their survival. This assessment appears to ignore the severe declines of the same species in agricultural areas of the Midwest (Herkert 1991), a region that presumably would be considered the center of abundance for most of these species.

Northeastern Grasslands before European Settlement

Although many popular writers and botanists have assumed that eastern North America was carpeted with a continuous forest before

European settlement (Day 1953), this view has been questioned by botanists since the early 1900s. Bromley (1935) argued that periodic burning by Native Americans created large grassland areas in southern New England, and Day (1953) presented extensive evidence that open habitats such as croplands and meadows were common in the pre-European settlement landscape. Day quoted the accounts of many early explorers and colonists who observed open landscapes created by firewood harvesting, clearing for maize fields, and burning to enhance hunting areas. For example, he cited Giovanni da Verrazano's account of the area around Narragansett Bay, RI, in 1524: open plains, without trees or forests, occurred 40 to 45 kilometers (km) inland from the bay. Verrazano, Samuel de Champlain, and John Smith all reported extensive areas of cleared land along the New England coast before Europeans colonized the area (Whitney 1994), and an early settler in Salem, MA, described "open plains, in some places five hundred acres...not much troublesome for to cleere for the plough to goe in" (Day 1953). Nor was open land restricted to coastal areas; Native Americans had also cleared inland river valleys for farming and hunting (Patterson and Sassaman 1988). In 1642 John Winthrop (*in* Hosmer 1959) described how the second European expedition to the White Mountains passed through "many thousands of acres of rich meadow" while paddling birch-bark canoes up the Saco River in what is now Maine.

Early assessments that Native American agriculture had relatively little effect on the landscape were based on postcontact population estimates. Yet population densities were much higher before contact with Europeans triggered massive epidemics that killed a high proportion of the people in most New England tribes (Crosby 1972, Cronon 1983, Denevan 1992, Whitney 1994). As Kulikoff (1986) writes regarding the Chesapeake Bay area, "though English settlers did not find a wilderness, they did create one": extensive agricultural clearings reverted to forest as Native American populations declined.

Pilgrims traveling through the area near Warren, MA, in 1621 "saw the remains of so many once occupied villages and such extensive formerly cultivated fields that they concluded thousands of people must have lived there before the plague" (Russell 1980). Early maps, drawings, and written accounts of the landscape around Native American settlements in the southeastern United States before European settlement provide direct evidence of extensive agricultural clearings and circumstantial evidence of parklands maintained by controlled burning (Hammett 1992). In New York and southern New England, relatively high population densities combined with the farming practices used by most of the tribes of this region would have resulted in extensive areas of cleared land in the form of both active and abandoned fields. Whitney (1994) argues that eastern Native Americans used slash-and-burn methods similar to those still used in many parts of the tropics: a plot would be used for a few years until fertility declined, then it would be abandoned and a new plot would be cleared. Patterson and Sassaman (1988) concluded that the use of slash-and-burn agriculture in southern New England resulted in a "mosaic of forests and fields in varying stages of succession." However, Doolittle (1992) points out that although there are descriptions of forest clearing by Native Americans, most descriptions of farming at the time of European contact indicate that Native Americans used large permanent fields from which tree stumps had been removed. These fields may have been taken out of production periodically to become the fallow "weed-covered" fields near Boston Bay described by Champlain in 1605 (Doolittle 1992).

Either slash-and-burn or permanent-field agriculture would have produced early-successional habitats that could have been used by grassland birds; the former would have continuously generated recently abandoned fields, whereas the latter would have required "resting" fallow fields.

There also is good evidence that the Native Americans of eastern North America burned large areas to create open woodlands and grassland for hunting. Although Russell (1983) argues that the evidence for extensive burning by Native Americans is weak, Whitney (1994) cites numerous detailed accounts of Native Americans using fire to clear vegetation. For example, colonist Roger Williams (1963) wrote that the tribes in New England "burnt up all the underwoods in the Countrey, once or twice a yeare and therefore as Noble men in England possessed great Parkes...onely for their game." And English settler William Wood (1634), describing spring in Massachusetts in the 1630s, wrote that "when the grass begins to put forth it grows apace, so that where it was all black by reason of the winter's burnings, in a fortnight there will be grass a foot tall." In another passage, Wood (1634) described how Native Americans used burning to "suppress the underwood, which else would grow all over the country." Thomas Morton (1637) described how Native Americans set fire to the country in both fall and spring, destroying the understory and stunting tree growth so that large trees could be found only "in the lower grounds where the grounds are wett when the country is fired." In 1818 B. Trumbull (*in* Olmsted 1937) reported that Native Americans in Connecticut "so often burned the country, to take deer and other small game, that in many of the plain dry parts of it, there was but little small timber. Where the lands were burned there grew bent grass...two, three and four feet high." Whitney (1994) describes additional first-

History of Grasslands in the Northeast

hand reports of Native Americans in the Northeast and Ontario using fire to create open lands.

Although fires were probably infrequent in most forests that were remote from Native American settlements (Russell 1983), fire and other disturbances near settlements provided extensive habitat for early-successional species, including grassland birds. An analysis of charcoal deposits in the sediments of 11 New England lakes indicates that before European settlement fires were frequent in densely populated coastal areas but infrequent in inland and northern areas (Patterson and Sassaman 1988). Winne (1988) provides a detailed analysis of pollen and charcoal in lake sediments for two watersheds (Pineo and Mud ponds) near the coast in eastern Maine. After a major fire destroyed an eastern hemlock (*Tsuga canadensis*) forest at Pineo Pond about 825 years ago, fires occurred in this watershed every 125 to 150 years, resulting in scrubby, fire-adapted vegetation. In contrast, fires burned the forest of the Mud Pond watershed every 230 to 300 years. Judging by the amount of charcoal deposited, the Mud Pond fires were more intense than those near Pineo Pond, but the shrubby vegetation generated by a fire near Mud Pond was replaced by forest before another fire occurred. The patterns in these two watersheds indicate that both temporary openings and more permanent open "barrens" were created by fire in eastern Maine. This view is supported by reports of surveyors who visited the area in 1792, soon after European settlement. They described the area north of Pineo Pond as "a large sandy plain [with] some scattering pines and birch bushes" and reported that Epping Plain, which was southeast of Pineo Pond, was a "clear level sandy plain which is 4 miles wide....capable of parading and maneuvering 40,000 Men" (Pierpont and Albee 1792). Today an open shrubland is maintained for lowbush blueberry (*Vaccinium angustifolium*) production in the Pineo Pond area by burning, mowing, and herbicide spraying. Several species of grassland birds nest in this habitat (Vickery et al. 1994).

Northeastern Grasslands before Native American Agriculture

According to Smith (1989), the patterns of Native American land use observed in the 1600s began to emerge about 2,000 years ago, with maize becoming the dominant crop only after A.D. 800. Perhaps many species of grassland birds colonized these cultivated areas after the initiation of Native American, rather than European, agriculture. If so, the current declines of these species would represent a return to conditions before humans began to modify the vegetation of eastern North America substantially.

Some of the apparently "natural" open grasslands of eastern North America are the product of human activities. For example, historical accounts and pollen analyses indicate that the extensive heathlands and sandplain grasslands on Nantucket Island, MA, resulted from the clearing of oak (*Quercus*) forest and grazing of sheep after Europeans settled that island (Dunwiddie 1989). Some of the open habitats on the East Coast, however, may predate disturbance by either Native Americans or Europeans. A good candidate for an ancient eastern grassland is the Hempstead Plains on Long Island, NY. Until it was developed for farms and housing after 1915 (Conrad 1935), the Hempstead Plains consisted of 24,300 hectares (ha) of rolling, essentially treeless prairie (Stalter and Lamont 1987). This grassland was described by travelers as early as 1670 (Harper 1911, Stalter and Lamont 1987). During most of the period of European settlement, the Plains was used primarily for sheep grazing and horse racing (Svenson 1936). Conrad (1935) provided a vivid description of the area in the early 1930s when relatively large tracts of prairie still existed. He reported that these "remain in essentially primeval condition showing in summer an expanse of yellow-green coarse bunch grass (*Schizachyrium scoparius* [little bluestem]), 1/2 [meter] m. tall dotted thickly with 1/2 m. hemispheres of *Baptista tinctoria* [wild indigo]....In May the prairie is blue with *Viola lineariloba* [*Viola pedata*; birdfoot violet] and *V. fimbriatula* [ovate-leaved violet]....In late June *Baptista* is full of yellow pea-shaped flowers and many clumps of silvery *Tephrosia virginiana* [goat's rue] show their heads of pink and yellow bloom." In addition to little bluestem, other species typical of tallgrass prairies of the Midwest, such as big bluestem (*Andropogon gerardii*),

History of Grasslands in the Northeast

birdfoot violet, and upland willow (*Salix humilis*), were common on the Hempstead Plains (Conrad 1935). The site was characterized by frequent fires and thin soil resting on a porous substratum of quartz and granite pebbles (Conrad 1935, Cain et al. 1937), features that may have favored the growth of grasses and herbs rather than trees and shrubs.

The Montauk Downs of western Long Island and the large glades of the Allegheny Mountains of Pennsylvania also were open grasslands at the time of European settlement (Whitney 1994). Moreover, a large savanna dominated by tall, wiry grass and scattered large oaks covered an extensive area of sandy soil along the east side of the Quinnipiac River north of New Haven, CT (Olmsted 1937). Later, after this area had been degraded by overgrazing, it was known as the North Haven Sand Plains. It is uncertain if such openings in the eastern forest were created by the activities of Native Americans.

Smaller shrubby and grassy openings in the eastern forest resulted from dam-building by beavers (*Castor canadensis*). After a beaver dam is abandoned, the pond behind the dam drains. This marshy area often becomes a "beaver meadow," a patch of shrubby vegetation or grassland. Although beaver ponds and meadows are largely restricted to floodplains, their ecological impacts were probably extensive before beavers were extirpated in most of their range in the Northeast by the nineteenth century (Naiman et al. 1988). Beaver activity may influence 20 to 40 percent of the total length of second- to fifth-order streams (Naiman et al. 1988). After beavers became reestablished at the Quabbin Reservation in Massachusetts in 1952, the population grew exponentially until the density reached 0.8 colony per kilometer of stream (Howard and Larson 1985). In Ontario's Algonquin National Park, where beaver populations are protected, there is a high density of beaver ponds and meadows (Coles and Orme 1983). In the Adirondack Mountains of New York, beaver dams created patches of disturbance that covered an average area of 6.7 ha, with a maximum area of 12 ha (Remillard et al. 1987). Analysis of the vegetation of 39 beaver sites in the Adirondack Mountains over a period of 40 years using aerial photographs indicated that particular sites go through a 10- to 30-year cycle of abandonment by beavers, succession through emergent, shrub-emergent, and shrub stages, and reestablishment of a beaver pond (Remillard et al. 1987).

The impact of a dense beaver population can be considerable. Bela Hubbard, an early land surveyor who surveyed land in Michigan before European settlement, described one-fifth of the area within 19 km of present-day Detroit as "marshy tracts or prairies which had their origin in the work of the beaver" (Whitney 1994). Coles and Orme (1983) argued that the forest of prehistoric England must have been "moth-holed with clearings wherever beaver were present." These "grassy meadows of relict pools" were also an important feature of the presettlement landscape of eastern North America.

Historical descriptions of presettlement vegetation and analysis of the regional frequency of fires, windstorms, and other disturbances indicate that some areas of the Northeast were characterized by an almost continuous forest canopy with few large openings (Borman and Likens 1979, Seischab and Orwig 1991). This was particularly true for northern-hardwood forests far from the coast. Other regions of the Northeast, however, were subject to disturbances that created large grassland openings in the forest (Runkle 1990).

Many bird species that are disappearing from the Northeast today may have been components of these eastern grassland communities before the arrival of Europeans and perhaps even before the advent of Native American agriculture. If so, then these bird communities may be a link to the bird communities of the spruce (*Picea*) parkland, a grassy savanna with scattered spruce trees that stretched in a wide band from the East Coast to the Great Plains between 18,000 and 12,000 years ago (Webb 1988). During this period, bird skeletons were deposited in owl pellets in caves at Natural Chimneys, VA (Guilday 1962). Although some woodland species such as Eastern Wood-Pewee (*Contopus virens*) and Red-bellied Woodpecker (*Melanerpes carolinus*) were present, the remains of many other species (Sharp-tailed Grouse [*Tympanuchus phasianellus*], Northern Bobwhite [*Colinus virginianus*], Upland Sandpiper, Red-headed Woodpecker [*Melanerpes erythrocephalus*], Black-billed Magpie [*Pica pica*], and Brown-headed Cowbird [*Molothrus ater*]) indicate that much of the landscape was open grassland and savanna.

Between 12,000 and 10,000 years ago, the spruce parkland in New England was replaced by forest (Webb 1988). At the beginning of this postglacial period, and through all of the earlier interglacial periods of the Pleistocene, large browsers such as mastodons may have maintained grassy openings in the forest or may even have converted forest into savanna in some regions, as African elephants do today (Andersson and Appelquist 1990, Puchkov 1992). Mastodons, ground sloths, and other giant herbivores disappeared from North America only in the present interglacial period (Martin and Klein 1984). Human activities such as burning and farming subsequently may have created grassy openings, permitting some grassland species to persist in forests despite the absence of open habitats created by giant browsing mammals (Andersson and Appelquist 1990).

The Origin of Northeastern Grassland Birds

The common impression that many grassland bird species spread eastward from the prairies of the Midwest to the newly cleared farmland of the East Coast is substantiated by several well-documented examples of range expansion. The prairie subspecies of the Horned Lark (*Eremophila alpestris praticola*), for instance, spread eastward from Illinois and Wisconsin, reaching Michigan and Ontario in the 1870s, New York in the 1880s, New England by 1891, and Pennsylvania and Maryland by 1910 (Forbush 1927, Thomas 1951, Hurley and Franks 1976). The Dickcissel (*Spiza americana*) spread eastward from the tallgrass prairies in the early 1800s, but its range contracted after 1850 and it eventually disappeared as a regular breeding bird along the East Coast (Hurley and Franks 1976). The Western Meadowlark (*Sturnella neglecta*) expanded its range into Wisconsin and Michigan after 1900 (Lanyon 1956), and the Lark Sparrow (*Chondestes grammacus*) spread from the prairies to agricultural areas in the Ohio Valley, West Virginia, and western Maryland (Brooks 1938).

Although these eastward range expansions were well documented, we do not have similar evidence for invasion of the Northeast by the species that are most abundant and widespread in eastern grasslands. Upland Sandpipers, Grasshopper Sparrows, Bobolinks (*Dolichonyx oryzivorus*), Eastern Meadowlarks (*Sturnella magna*), and other common grassland birds were reported by the earliest ornithologists who systematically documented the distribution of birds on the eastern coast of North America. Alexander Wilson's *American Ornithology* (originally published between 1808 and 1814; see Brewer 1839) and John James Audubon's *Ornithological Biography* (1831–1849) were published more than 100 years after most of the eastern seaboard had been cleared, so it is possible that grassland birds colonized the meadows and pastures created by Europeans long before their occurrences were initially documented. Some seventeenth-century European observers, such as John Josselyn (Lindholt 1988), described gamebirds and the more conspicuous songbirds, but only a few species are recognizable because the descriptions are sketchy and the authors used British bird names for North American species. Even if a diversity of grassland birds had been present, it would not be surprising if they were missing from these early accounts considering that many grassland birds are small, inconspicuous, and plainly marked.

The Heath Hen

Unlike other grassland birds, the now extinct Heath Hen (*Tympanuchus cupido cupido*) was described by the first European settlers in New England. This eastern subspecies of the Greater Prairie-Chicken was noticed because of its importance as a game bird. William Wood (1634) described "heathcocks and partridges" (presumably Heath Hens and Ruffed Grouse [*Bonasa umbellus*], respectively) as common in Massachusetts and indicated that a hunter could kill half a dozen in a morning. Thomas Morton (1637), a contemporary of Wood's, described Heath Hens as so common that hunters seldom wasted shot on them. Ornithologist Thomas Nuttall (1840) reported that this bird was "so common on the ancient bushy site of Boston, that laboring people or servants stipulated with their employers not to have Heath Hen brought to table oftener than a few times in the week." Perhaps not surprisingly, the Heath Hen became extinct in 1932 (Greenway 1958).

Heath Hens inhabited sandy scrub oak plains, blueberry barrens, grasslands, and other open habitats on the eastern seaboard between southern Maine and Virginia (Forbush 1927, Johnsgard 1983) or, more definitely, between Massachusetts and Maryland (Greenway 1958). According to Nuttall (1840), they were "confined to dry, barren, and bushy tracts, of small extent." Alexander Wilson (see Brewer 1839) described the favorite haunts of this species as "open, dry plains, thinly interspersed with trees, or partially overgrown with scrub oak," and in the early 1800s Audubon (1831–1849) wrote that Heath Hens were found in New Jersey, on the "brushy plains" of Long Island, NY, and in Massachusetts on Martha's Vineyard and the Elizabeth Islands. Other observers described the Heath Hen as occurring on the sandy plains of coastal New Jersey before the 1850s, and in "huckleberry [*Gaylussacia*] barrens" in Pennsylvania in the 1860s (Greenway 1958).

Most information on the diet of the Heath Hen comes from Martha's Vineyard, where conservationists made a concerted effort to save the last population after the mainland populations had become extinct by 1870 (Gross 1932). The diet of Heath Hens shifted from shoots of grasses and forbs in the spring to berries and insects in the summer and autumn and to acorns, seeds, and berries (especially bayberry [*Myrica pensylvanica*]) in the winter (Gross 1932). Acorns may have been particularly important for winter survival of both Heath Hens and the Great Plains subspecies of Greater Prairie-Chicken (*Tympanuchus cupido pinnatus*), and both subspecies may have been most abundant in landscapes with low vegetation and an abundance of oak (e.g., oak savanna, grassland with scrub oak, and interspersed patches of grassland and oak woodland; Johnsgard 1983, Schroeder and Robb 1993).

Ornithologist William Brewster classified the Heath Hen and the Greater Prairie-Chicken of the Great Plains as separate species, but they differ only slightly and were later classified as different subspecies of the same species (Gross 1932). Even the subspecific status of Heath Hen is now questioned (Schroeder and Robb 1993), but at the very least the Heath Hen represented a distinct regional population that was isolated from populations on the Great Plains. In contrast with the Great Plains birds, the eastern birds had thicker barring on the underparts, darker thighs, and fewer display feathers (pinnae) on the neck (Schroeder and Robb 1993). The distinctiveness of the isolated eastern population indicates that Greater Prairie-Chickens existed on the East Coast long before agriculture and artificial fires were introduced into the region. This, in turn, suggests that open habitats and grassland bird communities existed before the forests were cleared.

Heath Hens were common on Long Island, NY, in open scrub oak, pine (*Pinus*) barrens, and the open grasslands of the Hempstead Plains until the early nineteenth century (Brewer 1839). It is unknown whether they coexisted with other grassland birds, but Upland Sandpipers, Bobolinks, and Vesper and Grasshopper Sparrows were common on the Hempstead Plains in the early 1900s (Bull 1974). Perhaps these species lived alongside Heath Hens on these wide expanses of little bluestem prairie and in other open habitats along the East Coast before the arrival of Europeans.

Other Endemic Taxa in Northeastern Grasslands

Other species and subspecies are restricted to grasslands along the East Coast of North America, suggesting that these populations have existed in isolation in the East for many thousands of years, perhaps since unbroken grasslands reached from the Great Plains to the Atlantic as the glacial ice sheet melted. The Eastern Henslow's Sparrow (*Ammodramus henslowii susurrans*) has a breeding range restricted to central New York and southern New England south to Virginia, eastern West Virginia, and North Carolina (Smith 1968). Some plant species are also endemic to eastern grasslands (P. Dunwiddie pers. comm.): bushy rockrose (*Helianthemum dumosum*) is found from Massachusetts to Long Island; sandplain agalinis (*Agalinis acuta*) is found from eastern Massachusetts to Maryland; and sickle-leaved golden aster (*Pityopsis falcata* [= *Chrysopsis falcata*]) is found from Massachusetts to New Jersey (Gleason and Cronquist 1991). A variety of blazing star (northern blazing star, *Liatris scariosa* var. *novae-angliae*, previously called *L. borealis*) is restricted to grasslands in New York and New England.

Conservation of Grassland Birds

Because of the paucity of historical records of small birds from the period of European settlement, it is likely that only fossil or subfossil remains could provide definitive evidence of the occurrence of grassland birds in eastern North America before 1620. Given the strong evidence that extensive grasslands, savannas, and shrublands occurred in this region at the time of European settlement, however, it is reasonable to conclude that many open-country species are native to the region, not recent invaders from the western prairies. During the eighteenth and nineteenth centuries, these species undoubtedly became much more abundant than they had been before Europeans cleared the land, but later they may have declined far below the level of abundance characteristic of the presettlement landscape. Many species are now in danger of regional extinction. These species deserve the same attention from conservationists that birds associated with forests, marshes, and lakes have received.

Many of the original grasslands, such as beaver meadows and recently burned areas, were ephemeral. Some areas may have been disturbed frequently enough to create stable grasslands; the Hempstead Plains of Long Island and some of the barrens of Maine are obvious candidates. Temporary grasslands probably are created much less frequently today because beavers are less abundant and fires are controlled. Most of the areas that may have been stable grasslands have been developed for agriculture or housing. For example, except for two patches comprising less than 36 ha, the Hempstead Plains have been completely developed (Antenen et al. 1994). Other open areas, such as the North Haven Sand Plains in Connecticut, have suffered a similar fate. An exception is the blueberry barrens of eastern Maine, which have been maintained in a seminatural state by controlled burning to sustain commercial blueberry production (Vickery et al. 1994).

With the exception of Maine's blueberry barrens and a few other areas that have remained as seminatural open habitat for centuries, the habitats used by grassland birds in the Northeast are artificial. Grassland species have declined primarily because much of the farmland in the Northeast has been abandoned and has reverted to forest and because the remaining farmland is now managed more intensively for agricultural production (Askins 1993). For example, hayfields have become less suitable as nesting habitat for Eastern Meadowlarks, Bobolinks, and some other grassland species because they are mowed earlier in the summer (before the end of the nesting season) and because they are rotated more frequently (Bollinger and Gavin 1992). As much of the farmland once used by grassland species has disappeared or became unsuitable for nesting, other types of artificial grasslands have became increasingly important for sustaining regional populations. For example, most of the remaining populations of Grasshopper Sparrows and Upland Sandpipers in southern New England are found in extensive mowed areas at airports and military airfields (Veit and Petersen 1993, Bevier 1994, Melvin 1994).

Regional populations of grassland birds can be maintained with proper management of artificial grasslands such as hayfields, fallow farmland, and mowed areas near airport runways. Farmland, for example, can be managed for grassland birds and early-successional plants and insects through the Conservation Reserve Program, which pays farmers to take land out of production and manage it for soil and wildlife conservation (Frawley and Best 1991, Dunn et al. 1993). Relatively simple changes in airport management (e.g., removal of woody vegetation and changes in mowing schedules to avoid the nesting season) have sustained or improved habitat for grassland birds at Westover Air Reserve Base in Massachusetts (Melvin 1994), Bradley International Airport in Connecticut (Crossman 1989), and Floyd Bennett Field, a former naval air base, on Long Island, NY (Lent and Litwin 1989). Habitat management at Westover Air Reserve Base resulted in substantial increases in the abundance of nesting Grasshopper Sparrows and Upland Sandpipers between 1987 and 1994 (Melvin 1994).

Successful management of grassland birds depends on understanding their habitat requirements. Many species are habitat specialists (Askins 1993). Horned Larks, for example, tend to occur in open areas with sparse vegetation (Whitmore and Hall 1978), Grasshopper Sparrows in areas dominated by bunchgrass with patches of bare ground (Whitmore 1981), and Henslow's Sparrows in areas dominated by tall dense grass with little or no bare ground (Zimmerman 1988). Although recently burned or mowed grassland provides favorable habitat for Horned Larks and Grasshopper Sparrows, Henslow's Sparrows need grassland that has not been disturbed for several years. Only a mosaic of grassland patches in different stages of recovery from disturbance will support all of these species (Renken and Dinsmore 1987, Zimmerman 1988).

Another important feature of habitat quality for many grassland specialists is the area of continuous grassland habitat. In Illinois (Herkert 1994a) and Maine (Vickery et al. 1994), both the density and diversity of grassland birds increased with habitat area. In Illinois, few Grasshopper Sparrows, Savannah Sparrows (*Passerculus sandwichensis*), Bobolinks, or Henslow's Sparrows were found on patches of grassland smaller than 30 ha (Herkert 1994a). In Maine, Grasshopper Sparrows were absent on sites with less than 30 ha of grassland (Vickery et al. 1994). These results indicate that most of the effort to maintain populations of grassland birds should be directed at extensive areas of grassland. A large grassland not only can accommodate species that are absent or less common in smaller grassland patches, but it can also be managed as a habitat mosaic that provides different types of habitat for different species, which is not feasible on small grassland patches (Herkert 1994b).

Expending scarce resources to maintain hayfields, fallow fields, and airfields may seem unwise to conservationists accustomed to protecting forests and wilderness areas. Yet a large number of bird, insect, plant, and other species depend on these grassland habitats. Artificial habitats are critical for many of these species because people have destroyed most of the native grassland habitat, including most of the midwestern prairies where these species were once most abundant (Bollinger et al. 1990). People also have interrupted or dampened many of the natural processes of disturbance, such as fires and beaver activity, that once created the early-successional habitats that grassland species need. Looking to the near future, artificial grasslands represent our best hope for maintaining grassland species in the northeastern United States. These species are an important, and probably an ancient, component of biological diversity in the region.

Acknowledgments

I have gained insights about the questions addressed in this paper in discussions with P. Vickery, W. Niering, G. Dreyer, and R. DeGraaf. P. Vickery and J. Herkert gave me useful advice on the manuscript. My research was supported by funds provided by the U.S. Department of Agriculture Forest Service, Northeastern Forest Experiment Station. A revised version of this paper will appear in a forthcoming book to be published by Yale University Press.

Literature Cited

Andersson, L., and T. Appelquist. 1990. The influence of the Pleistocene megafauna on the nemoral and the boreonemoral ecosystem: a hypothesis with implications for nature conservation strategy. Svensk Botanisk Tidskrift 84: 355–368.

Antenen, S., M. Jordan, K. Motivans, J. B. Washa, and R. Zaremba. 1994. Hempstead Plains fire management plan, Nassau County, Long Island, New York. Report to Long Island Chapter of The Nature Conservancy, Cold Spring Harbor, NY.

Askins, R. A. 1993. Population trends in grassland, shrubland, and forest birds in eastern North America. Pp. 1–34 *in* Current ornithology, vol. 11 (D. M. Power, ed.). Plenum Press, New York.

Askins, R. A., J. F. Lynch, and R. Greenberg. 1990. Population declines in migratory birds in eastern North America. Pp. 1-57 *in* Current ornithology, vol. 7 (D. M. Power, ed.). Plenum Press, New York.

Audubon, J. J. 1831-1849. Ornithological biography, or, an account of the habits of the birds of the United States of America. Vol. 1, J. Dobson, Philadelphia; vol. 2-5, A. and C. Black, Edinburgh.

Bevier, L. R., ed. 1994. The atlas of breeding birds of Connecticut. State Geol. Nat. Hist. Survey Conn. Bull. 113.

Bollinger, E. K., P. B. Bollinger, and T. A. Gavin. 1990. Effects of hay-cropping on eastern populations of the Bobolink. Wildl. Soc. Bull. 18: 142-150.

Bollinger, E. K., and T. A. Gavin. 1992. Eastern Bobolink populations: ecology and conservation in an agricultural landscape. Pp. 497-506 *in* Ecology and conservation of neotropical migrant landbirds (J. M. Hagan III and D. W. Johnston, eds.). Smithsonian Inst. Press, Washington, D.C.

Borman, F. H., and G. E. Likens. 1979. Catastrophic disturbances and the steady state in northern hardwood forests. Am. Sci. 67: 660-669.

Brewer, T. M. 1839. Wilson's American ornithology. Charles L. Cornish, New York.

Bromley, S. W. 1935. The original forest types of southern New England. Ecol. Monogr. 5: 61-89.

Brooks, M. 1938. The Eastern Lark Sparrow in the upper Ohio valley. Cardinal 4: 181-200.

Bull, J. 1974. Birds of New York state. Doubleday, Garden City, NY.

Cain, S. A., M. Nelson, and W. McLean. 1937. Andropogonetum Hempsteadi: a Long Island grassland vegetation type. Am. Midl. Nat. 18: 334-350.

Coles, J. M., and B. J. Orme. 1983. *Homo sapiens* or *Castor fiber*? Antiquity 57: 95-102.

Conrad, H. S. 1935. The plant associations of central Long Island. A study in descriptive plant sociology. Am. Midl. Nat. 16: 433-516.

Cronon, W. 1983. Changes in the land. Indians, colonists, and the ecology of New England. Hill and Wang, New York.

Crosby, A. W., Jr. 1972. The Columbian exchange: biological and cultural consequences of 1492. Greenwood Press, Westport, CT.

Crossman, T. I. 1989. Habitat use of Grasshopper and Savannah sparrows at Bradley International Airport and management recommendations. Master's thesis, Univ. of Connecticut, Storrs.

Day, G. M. 1953. The Indian as an ecological factor in the northeastern forest. Ecology 34: 329-346.

Denevan, W. M. 1992. The pristine myth: the landscape of the Americas in 1492. Ann. Assoc. Am. Geogr. 82: 369-385.

Doolittle, W. E. 1992. Agriculture in North America on the eve of contact: a reassessment. Ann. Assoc. Am. Geogr. 82: 386-401.

Dunn, C. P., F. Stearns, G. R. Guntenspergen, and D. M. Sharpe. 1993. Ecological benefits of the Conservation Reserve Program. Conserv. Biol. 7: 132-139.

Dunwiddie, P. W. 1989. Forest and heath: the shaping of vegetation on Nantucket Island. J. Forest Hist. 33: 126-133.

Forbush, E. H. 1927. Birds of Massachusetts and other New England states. Vol. 2. Mass. Dept. Agric., Boston.

Frawley, B. J., and L. B. Best. 1991. Effects of mowing on breeding bird abundance and species composition in alfalfa fields. Wildl. Soc. Bull. 19: 135-142.

Gleason, H. A., and A. Cronquist. 1991. Manual of vascular plants of northeastern United States and adjacent Canada. 2d ed. N. Y. Botanical Garden, Bronx, NY.

Greenway, J. G. 1958. Extinct and vanishing birds of the world. Spec. Publ. no. 13, Am. Comm. Internatl. Wild Life Protection, New York.

Gross, A. O. 1932. Heath Hen. Pp. 264-280 *in* Life histories of North American gallinaceous birds (A. C. Bent, ed.). U.S. Natl. Mus. Bull. 162.

Guilday, J. E. 1962. The Pleistocene local fauna of the Natural Chimneys, Augusta County, Virginia. Ann. Carnegie Mus. 36: 87-122.

Hammett, J. E. 1992. The shapes of adaptation: historical ecology of anthropogenic landscapes in the southeastern United States. Landscape Ecol. 7: 121-135.

Harper, R. M. 1911. The Hempstead Plains: a natural prairie on Long Island. Bull. Am. Geog. Soc. 43: 351-360.

Herkert, J. R. 1991. Prairie birds of Illinois: population response to two centuries of habitat change. Ill. Nat. Hist. Surv. Bull. 34: 393-399.

Herkert, J. R. 1994a. The effect of habitat fragmentation on midwestern grassland bird communities. J. Ecol. Appl. 4: 461-471.

Herkert, J. R. 1994b. Breeding bird communities of midwestern prairie fragments: the effects of prescribed burning and habitat-area. Nat. Areas J. 14: 128-135.

Hosmer, J. K. 1959. Winthrop's journal, "History of New England," 1630-1649. Barnes and Noble, New York.

Howard, R. J., and J. S. Larson. 1985. A stream habitat classification system for beaver. J. Wildl. Manag. 49: 19-25.

Hurley, R. J., and E. C. Franks. 1976. Changes in the breeding ranges of two grassland birds. Auk 93: 108-115.

James, F. C., D. A. Wiedenfeld, and C. E. McCulloch. 1992. Trends in breeding populations of warblers: declines in the southern highlands and increases in the lowlands. Pp. 43-56 *in* Ecology and conservation of neotropical migrant landbirds (J. M. Hagan III and D. W. Johnston, eds.). Smithsonian Inst. Press, Washington, D.C.

Johnsgard, P. A. 1983. The grouse of the world. Univ. of Nebraska Press, Lincoln.

Kulikoff, A. 1986. Tobacco and slaves: the development of southern cultures in the Chesapeake, 1680-1800. Univ. of N. Carolina Press, Chapel Hill.

Lanyon, W. E. 1956. Ecological aspects of the sympatric distribution of meadowlarks in the north-central states. Ecology 37: 98-108.

Lent, R. A., and T. S. Litwin. 1989. Bird-habitat relationships as a guide to ecologically-based management at Floyd Bennett Field, Gateway National Recreation Area. Part 1. Baseline study. Seatuck Research Prog., Cornell Univ. Lab. Ornithol., Islip, NY.

Lindholt, P. J. 1988. John Josselyn, colonial traveler: a critical edition of an account of two voyages to New-England. University Press of New England, Hanover, NH.

Martin, P. S., and R. G. Klein, eds. 1984. Quaternary extinctions. A prehistoric revolution. Univ. of Arizona Press, Tucson.

Melvin, S. 1994. Military bases provide habitat for rare grassland birds. Mass. Div. Fish. Wildl. Nat. Heritage News 4: 3.

Morton, T. 1637. New English Canaan. Reprint. De Capo Press, New York, 1969.

Naiman, R. J., J. M. Melillo, and J. E. Hobbie. 1988. Ecosystem alteration of boreal forest streams by beaver (*Castor canadensis*). Ecology 67: 1254-1269.

Nuttall, T. 1840. A manual of ornithology. 2d ed. Hilliard, Gray and Co., Boston.

Olmsted, C. E. 1937. Vegetation of certain sand plains of Connecticut. Bot. Gazette 99: 209-300.

Patterson, W. A., III, and K. E. Sassaman. 1988. Indian fires in the prehistory of New England. Pp. 107-135 *in* Holocene human ecology in northeastern North America (G. P. Nichols, ed.). Plenum Press, New York.

Peterjohn, B. 1994. The North American breeding bird survey. Birding 26: 386-398.

Pierpont, J., and W. Albee. 1792. A journal over 1,000000 acres of land in the counties of Hancock and Washington. Handwritten journal. Historical Soc., Philadelphia, PA.

Puchkov, P. V. 1992. Uncompensated Wuermian extinctions. Part 2. Transformation of the environment by giant herbivores. Vestnik Zoologii 0 (1): 58-66.

Remillard, M. M., G. K. Gruendling, and D. J. Bogucki. 1987. Disturbance by beaver (*Castor canadensis* Kuhl) and increased landscape heterogeneity. Pp. 103-122 *in* Landscape heterogeneity and disturbance (M. G. Turner, ed.). Springer Verlag, New York.

Renken, R. B., and J. J. Dinsmore. 1987. Nongame bird communities on managed grasslands in North Dakota. Can. Field-Nat. 101: 551-557.

Runkle, J. R. 1990. Gap dynamics in an Ohio *Acer-Fagus* forest and speculations on the geography of disturbance. Can. J. Forest Res. 20: 632-641.

Russell, E. W. B. 1983. Indian-set fires in the forests of the northeastern United States. Ecology 64: 78-88.

Russell, H. S. 1980. Indian New England before the Mayflower. University Press of New England, Hanover, NH.

Schroeder, M. A., and L. A. Robb. 1993. Greater Prairie-Chicken. Birds of North America no. 36 (A. Poole, P. Stettenheim, and F. Gill, eds.). Acad. Nat. Sci. and Am. Ornithol. Union, Philadelphia, PA.

Seischab, F. K., and D. Orwig. 1991. Catastrophic disturbances in the presettlement forests of western New York. Bull. Torrey Bot. Club 114: 330-335.

Smith, B. D. 1989. Origins of agriculture in eastern North America. Science 246: 1566-1571.

Smith, C. R., D. M. Pence, and R. J. O'Connor. 1993. Status of neotropical migratory birds in the northeast: a preliminary assessment. Pp. 172-188 *in* Status and management of neotropical migratory birds (D. M. Finch and P. W. Stangel, eds.). Gen. Tech. Rep. RM-229, U.S. Forest Serv., Rocky Mtn. Forest and Range Exper. Stn., Forth Collins, CO.

Smith, W. P. 1968. Eastern Henslow's Sparrow. Pp. 776-778 *in* Life histories of North American cardinals, grosbeaks, buntings, towhees, finches, sparrows, and allies (A. C. Bent, ed.). U.S. Natl. Mus. Bull. 237.

Stalter, R., and E. E. Lamont. 1987. Vegetation of Hempstead Plains, Mitchell Field, Long Island, New York. Bull. Torrey Bot. Club 114: 330-335.

Svenson, H. K. 1936. The early vegetation of Long Island. Brooklyn Bot. Garden J. 25: 207-227.

Terborgh, J. 1989. Where have all the birds gone? Princeton Univ. Press, Princeton, NJ.

Thomas, E. S. 1951. Distribution of Ohio animals. Ohio J. Sci. 51: 153-167.

Veit, R. R., and W. R. Petersen. 1993. Birds of Massachusetts. Massachusetts Audubon Soc., Lincoln.

Vickery, P. D. 1992. A regional analysis of endangered, threatened, and special concern birds in the northeastern United States. Trans. N. E. Sect. Wildl. Soc. 48: 1-10.

Vickery, P. D., M. L. Hunter, Jr., and S. M. Melvin. 1994. Effects of habitat area on the distribution of grassland birds in Maine. Conserv. Biol. 8: 1087-1097.

Webb, T., III. 1988. Eastern North America. Pp. 385-414 *in* Vegetation history (B. Huntley and T. Webb III, eds.). Kluwer Academic Publ., Hingham, MA.

Whitcomb, R. F. 1987. North American forests and grassland: biotic conservation. Pp. 163-176 *in* Nature conservation: the role of remnants of native vegetation (D. A. Saunders, G. W. Arnold, A. A. Burbidge, and A. J. M. Hopkins, eds.). Surrey Beatty & Sons, Chipping Norton, New South Wales, Australia.

Whitmore, R. C. 1981. Structural characteristics of Grasshopper Sparrow habitat. J. Wildl. Manage. 45: 811-814.

Whitmore, R. C., and G. A. Hall. 1978. The response of passerine species to a new resource: reclaimed surface mines in West Virginia. Am. Birds 32: 6-9.

Whitney, G. G. 1994. From coastal wilderness to fruited plain. Cambridge Univ. Press, New York.

Williams, R. 1963. The complete writings of Roger Williams. Vol. 2. Russell and Russell, New York.

Winne, J. C. 1988. History of vegetation and fire on the Pineo Ridge blueberry barrens in Washington County, Maine. Master's thesis, Univ. of Maine, Orono.

Wood, W. 1634. New England's prospect. Reprint, ed. A. T. Vaughan. Univ. of Massachusetts Press, Amherst, 1977.

Zimmerman, J. L. 1988. Breeding season habitat selection by the Henslow's Sparrow (*Ammodramus henslowii*) in Kansas. Wilson Bull. 100: 17-24.

Robert A. Askins: Department of Zoology, Connecticut College, New London, CT 06320.

Effects of Habitat Area on the Distribution of Grassland Birds in Maine

Peter D. Vickery, Malcolm L. Hunter, Jr., and Scott M. Melvin

Abstract

We used multiple and logistic regression analysis to study the breeding-area requirements of 10 species of grassland and early-successional birds at 90 grassland-barren sites in Maine. The incidence of six of the species was clearly sensitive to the area of grassland. Upland Sandpipers (*Bartramia longicauda*), the species with the largest area requirements, were infrequent at sites smaller than 50 hectares (ha) and reached 50 percent incidence at those of about 200 ha. Grasshopper Sparrows (*Ammodramus savannarum*) reached 50 percent incidence at about 100 ha, Vesper Sparrows (*Pooecetes gramineus*) at about 20 ha, and Savannah Sparrows (*Passerculus sandwichensis*) at about 10 ha. Incidence for three edge species, Brown Thrasher (*Toxostoma rufum*), Common Yellowthroat (*Geothlypis trichas*), and Song Sparrow (*Melospiza melodia*), was negatively correlated with open area, and incidence for Field Sparrows (*Spizella pusilla*) was not strongly influenced by grassland size. These results indicate that grasslands need to be about 200 ha in area if they are likely to support a diverse grassland bird fauna, and probably need to be considerably larger if they are likely to provide sufficient habitat for the grassland raptors Northern Harrier (*Circus cyaneus*) and Short-eared Owl (*Asio flammeus*). However, a random sample of 100 hayfields in Maine revealed that only 1 percent were larger than 64 ha. Conservation efforts seeking to protect rare grassland birds need to consider sites of at least 100 ha, and preferably 200 ha, and these are notably rare in Maine and probably throughout New England and eastern North America.

This manuscript is adapted from an earlier version that appeared in Conservation Biology, pp. 1087–1097, vol. 8, no. 4, December 1994. Reprinted by permission of Blackwell Science, Inc., Cambridge, MA.

Introduction

The concept that some species are area-dependent for their preferred habitat (Bond 1957) has prompted considerable interest among conservation biologists because of increasing concern about the effectiveness of nature reserves to support self-sustaining populations of area-dependent species (Shafer 1990). Similarly, in North America there is concern that forest fragmentation may contribute to the decline of neotropical migrant birds (Robbins 1979, Robbins et al. 1989). Although area requirements of forest birds have received intensive study in recent years (Robbins et al. 1989), few studies have examined patterns of area-dependence among inhabitants of early-successional ecosystems, especially in regions dominated by late-successional forests (Bollinger and Gavin 1992, Smith and Smith 1992, Rudnicky and Hunter 1993).

This has important conservation implications, especially in eastern North America, because (1) many species of grassland birds are declining throughout most of their ranges, particularly in the northeastern United States (Robbins et al. 1986, Askins 1993); (2) grasslands constitute the habitat with the largest number of birds listed by the state as endangered, threatened, or of special concern in New England and New York (Vickery 1992); and (3) grassland area in New England and New York has declined by 60 percent during the past 60 years (U.S. Dept. Agric. 1936–1991). Lack of basic information on area requirements for early-successional species such as grassland birds will compromise conservation efforts for these declining species.

It is clear that native grassland ecosystems persisted on the northeastern North American landscape well before pre-Colonial settlement (Askins 1993). For example, stratigraphic palynological charcoal evidence indicates that some sites in eastern Maine have been dominated by fire-adapted plants, including a high proportion of graminoids, for at least 800 years (Winne 1988). Because of the geomorphology and exposure of these sites, the area of these pre-Colonial grassland-barrens was probably large and much as it is today; "a plain two or three miles in diameter....The soil is perfectly barren and covered with a short kind of heath and no wood. It has the appearance of having been burned...The nature of the whole is singular and different from anything I ever saw" (A. Baring 1796 *in* Fischer 1954).

We sought to determine patterns of habitat occupancy of early-successional birds breeding at 90 grassland-barren sites in Maine, a state with about 90 percent forest cover (Ferguson and Kingsley 1972). The area of these sites ranged from less than 2 to more than 400 ha. Most sites occurred on well-drained glaciomarine deltas that were composed of a dry sandplain grassland vegetation dominated by native graminoids, forbs, and shrubs (Vickery et al. 1992). In the absence of wildfires, these sites have remained open as a result of traditional mowing and burning practices used for commercial lowbush blueberry (*Vaccinium angustifolium*) production.

Study Sites and Methods

Ninety study sites were located in a band 350 kilometers (km) long along the Maine coast, from Wells (43°23' N, 70°39' W) in York County to Machias (44°42' N, 67°29' W) and Alexander (45°05' N, 67°30' W) in Washington County (Fig. 1). We used a planimeter to measure recent U.S. Soil Conservation Service aerial photographs (scale 1:20,000, 1:15,840, 1:1,320) to determine the area of each site.

Nearly all sites, which ranged from less than 2 to more than 400 ha of continuous grassland-barren, were currently, or had been recently, managed for commercial blueberry production, which involves biennial mowing and burning. In recent years many of these grassland-barrens have been sprayed with herbicide to increase blueberry production. This practice dramatically reduces graminoid and forb cover (Yarborough and Bhowmik 1988). Vegetation on these sites was predominantly lowbush blueberry, with lesser amounts of graminoids (little bluestem [*Schizachyrium scoparium*], poverty grass [*Danthonia*], and sedge [*Carex*]), shrubs (chokeberry [*Photinia*, formerly *Aronia*], alder [*Alnus*], and birch [*Betula*]), and forbs (northern blazing star [*Liatris scariosa* var. *novae-angliae*] and whorled loosestrife [*Lysimachia quadrifolia*]).

Each grassland-barren was censused on randomly selected 100-meter (m) fixed-radius circular plots (Edwards et al. 1981, Verner 1985) whose centers were at least 50 m from any forest edge (except at sites in the 0–2-, 2–4-, and 4–8-ha classes). The number of circular census plots varied depending on the area of the barren, from 1 on sites smaller than 8 ha to 15 on a site larger than 200 ha. On sites with more than one census plot, plot centers were separated by more than 250 m. Each circular plot was censused for 5 minutes on two different occasions. Censuses were conducted between 0600 and 1000 hr from 1 June to 18 July, in either 1989 or 1990. To reduce the bias of censusing at different periods of the breeding season, the first census for each site was conducted 1-20 June and the second census 21 June-18 July of the same year. The total number of species, the number of individuals of each species, and the total number of individuals were recorded for each circular plot.

Sixteen species were found on these grassland-barrens (Table 1). Northern Harrier, Killdeer (*Charadrius vociferus*), Alder Flycatcher

(*Empidonax alnorum*), Horned Lark (*Eremophila alpestris*), Indigo Bunting (*Passerina cyanea*), and Lincoln's Sparrow (*Melospiza lincolnii*) were found on fewer than five census plots and were included only in estimates of total species and total number of

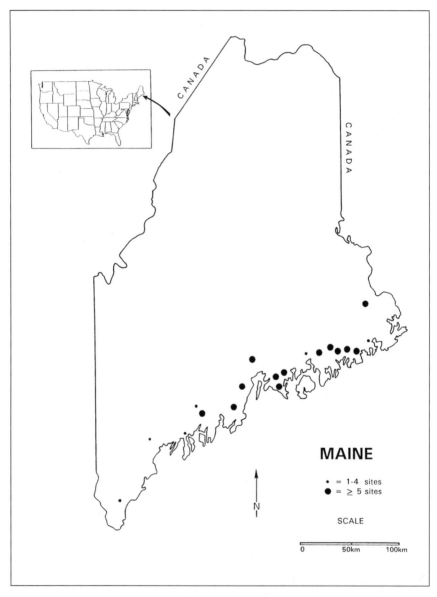

Figure 1. Distribution of grassland-barren sites censused in coastal Maine, 1989–1990.

Table 1. Bird species found on grassland-barrens in coastal Maine, 1989–1990.

Northern Harrier (*Circus cyaneus*)	Vesper Sparrow (*Pooecetes gramineus*)
Killdeer (*Charadrius vociferus*)	Savannah Sparrow (*Passerculus sandwichensis*)
Upland Sandpiper (*Bartramia longicauda*)	Grasshopper Sparrow (*Ammodramus savannarum*)
Alder Flycatcher (*Empidonax alnorum*)	Song Sparrow (*Melospiza melodia*)
Horned Lark (*Eremophila alpestris*)	Lincoln's Sparrow (*Melospiza lincolnii*)
Brown Thrasher (*Toxostoma rufum*)	Bobolink (*Dolichonyx oryzivorus*)
Common Yellowthroat (*Geothlypis trichas*)	Eastern Meadowlark (*Sturnella magna*)
Indigo Bunting (*Passerina cyanea*)	
Field Sparrow (*Spizella pusilla*)	

individuals. Grasshopper Sparrows reach the northeastern limit of their breeding range in Brunswick (43°54' N, 69°46' N), Cumberland Co. (Vickery 1990); all analyses for this species were therefore limited to sites (n = 10) within its range.

We used the line-transect technique to measure vegetative cover (Canfield 1941). A 100-m tape was unwound from the center to the edge of each census plot. Initial direction of the first transect line was determined randomly; the subsequent transects were 90, 180, and 270 degrees from the first. The dominant cover type—bare ground, stone, graminoid, forb, blueberry, low shrub (0–60 centimeters [cm]), high shrub (60 cm–2 m), or tree (more than 2 m)—was recorded in 50-cm units. In addition to determining which vegetation features might be associated with a species' presence, this technique permitted us to examine the potential importance of patchiness to these species. We calculated an index of patchiness by summing the number of changes (in 50-cm units) in cover type for each census point (maximum possible value: 4 transects x 200 possible units = 800). Percent cover for each habitat variable was arc-sine transformed. To normalize these data, the number of patches for each point-count and the area of each site were transformed to their natural logarithm.

We used multiple linear regression as a preliminary analysis to determine the importance of area in conjunction with vegetation structure and patchiness. To gain greater insight into how area affected incidence, we used logistic regression to generate incidence functions to predict the probability of a species' presence on a grassland-barren of a given size (Diamond 1975, Robbins et al. 1989, Wilkinson 1990). We used the linear-logistic model:

$$P(y = 1) = \frac{\exp(B_0 + B_1 x_1)}{1 + \exp(B_0 + B_1 x_1)}, \text{ where } x = \text{area.}$$

We then plotted the predicted probabilities of occurrence generated by the logistic regression analyses for the range of areas sampled (see Robbins et al. 1989, Wilkinson 1990 for further explanation). Because the large sample size in this study ($n = 90$) was normally distributed, we used the Z-test to determine the significance of each logistic regression coefficient for each species (Zar 1984). To eliminate the bias associated with uneven censusing effort (Haila 1986), only one randomly selected census plot was used to generate area-incidence functions for each species. We concur with Robbins et al. (1989) that 50 percent incidence provides a conservative but reasonable estimate of a species' minimum area requirements, and we have used this level to determine minimum requirements for grassland birds in this study.

To examine the potential relationship between bird species richness and grassland-barren size, we grouped sites into the following size classes (sample size in parentheses): 0 to 2 (15), 2 to 4 (9), 4 to 8 (15), 8 to 16 (14), 16 to 32 (14), 32 to 64 (6), and more than 64 (17) ha. Mean (±1 SE) number of species per size class was calculated from a randomly selected census plot from each grassland-barren. We used Spearman rank correlation to test (1) whether species richness changed as a function of area, and (2) to determine if the number of individuals per circular plot was associated with the number of species found on each circular plot (Wilkinson 1990).

To compare the size distribution of hayfields and pastures to grassland-barrens, we randomly sampled 100 hayfields or pastures and 100 grassland-barrens from aerial photographs, measured their area, and assigned them to the previously described size classes. We used a log-likelihood goodness of fit G-test to determine if these two types of open habitat differed in their size class distributions (Sokal and Rohlf 1981).

Results and Discussion

Multiple regression analysis demonstrated the importance of area for all 10 species (Table 2). The presence of seven species was positively correlated with increasing area. Three edge species—Common Yellowthroat, Brown Thrasher, and Song Sparrow—were negatively correlated with grassland-barren area. Habitat patchiness was an important feature for four of the seven species that were positively correlated with area, but it was unimportant for any of the edge species (Table 2). Habitat patchiness may be especially important for Upland Sandpipers; they demonstrated no other habitat associations but are known to require a variety of vegetative features for different parts of the breeding cycle (Ailes 1980), and in Illinois they do not occupy areas of uniform graminoid or forb cover (Buhnerkempe and Westemeier 1988). For all species, "year effect" (testing for differences between years) was not significant ($p > 0.40$).

Table 2. Variables identified as significant predictors of relative abundance of 10 species of grassland or edge birds in coastal Maine, 1989-1990.

Species	Number of Points at which Species Was Detected	Significant Predictors[a]	R^2	F[b]
Positive Area Effect				
Upland Sandpiper	39	area ($p < 0.001$) patchiness ($p < 0.001$)	0.190	13.3
Field Sparrow	15	area ($p < 0.05$) patchiness ($p < 0.001$) bare ground (−) ($p < 0.05$) high shrub ($p < 0.01$)	0.225	16.7
Vesper Sparrow	91	area ($p < 0.001$) patchiness ($p < 0.001$) bare ground ($p < 0.01$)	0.129	11.4
Savannah Sparrow	158	area ($p < 0.001$) bare ground (−) ($p < 0.001$) blueberry (−) ($p < 0.001$) high shrub (−) ($p < 0.001$)	0.219	16.2
Grasshopper Sparrow	14	area ($p < 0.05$) graminoid ($p < 0.001$) forb ($p < 0.01$)	0.484	12.0
Bobolink	36	area ($p < 0.001$) patchiness ($p < 0.001$) graminoid ($p < 0.01$) forb ($p < 0.001$)	0.369	26.8
Eastern Meadowlark	24	area ($p < 0.001$) graminoid ($p < 0.001$) low shrub ($p < 0.05$) high shrub (−) ($p < 0.05$)	0.259	16.0
Negative Area Effect				
Brown Thrasher	8	area (−) ($p < 0.05$) tree ($p < 0.001$)	0.167	23.2
Common Yellowthroat	27	area (−) ($p < 0.05$) high shrub ($p < 0.001$) tree ($p < 0.001$)	0.307	34.0
Song Sparrow	28	area (−) ($p < 0.001$) high shrub ($p < 0.001$) forb ($p < 0.001$)	0.483	71.9

[a] Significant predictors (variables): area, patchiness, bare ground, litter, graminoid, forb, stone, low shrub, high shrub, blueberry, tree, year. (−) indicates a significant negative correlation.

[b] $p < 0.001$ for all analyses.

Species with Positive Area Effects

The incidence functions for 6 of the 10 species showed positive correlations with increasing area (Fig. 2). Upland Sandpiper area requirements were the greatest; this species was rare on sites smaller than 50 ha, increased steadily with area ($Z = 3.08$, $p < 0.001$), and reached 50 percent incidence at about 200 ha. In Missouri, Upland Sandpiper incidence was 100 percent on plots larger than 100 ha (Samson 1980a), and in Illinois the species did not occur on plots smaller than 20 ha but reached an asymptote on plots in the 20- to 100-ha class (Herkert 1991a).

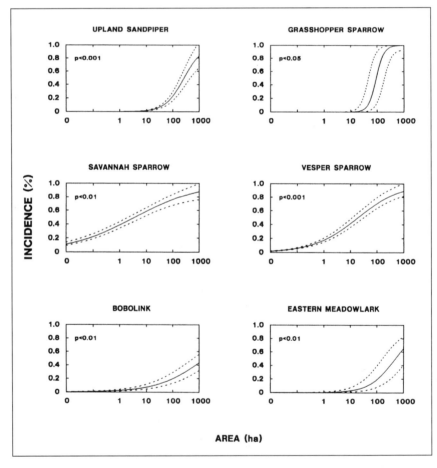

Figure 2. Incidence of six species of grassland birds was positively correlated with area of site. Dashed lines indicate 95 percent confidence intervals for the predicted probabilities.

Vesper Sparrow incidence increased with area and reached 50 percent at about 20 ha ($Z = 3.53$, $p < 0.001$; Fig. 2). Savannah Sparrow incidence increased with area and reached 50 percent at about 10 ha ($Z = 2.51$, $p < 0.01$). In Illinois, Herkert (1991b) found similar results; the minimum area for Vesper Sparrows was less than 10 ha, but for Savannah Sparrows it was 10 to 30 ha.

Incidence rose sharply for Grasshopper Sparrows and reached 50 percent at about 100 ha ($Z = 2.41$, $p < 0.01$; Fig. 2). In Wisconsin, Grasshopper Sparrows were more likely to occupy large (130–486 ha) rather than small (16–32 ha) tallgrass prairie fragments (Johnson and Temple 1986). Both our findings and those of Johnson and Temple (1986) suggest a larger minimum area for Grasshopper Sparrows than was found in Missouri (< 20 ha; Samson 1980a) and Illinois (10–30 ha; Herkert 1991b). The difference in Maine may be related to the relative rarity of this species in Maine and New England. As general population levels diminish, Grasshopper Sparrows are probably more selective in their area and habitat requirements, as has been shown for many species of birds (O'Connor 1981).

Although Bobolinks (*Dolichonyx oryzivorus*) and Eastern Meadowlarks (*Sturnella magna*) showed positive area effects (Fig. 2), the low incidence for these two species (< 40 percent at 500 ha) reflects the fact that many of the sites in this study were unsuitable for these species. Because both of these species prefer sites with extensive graminoid cover (Lanyon 1957, Bollinger and Gavin 1992), and were infrequent in sites that had been recently sprayed with herbicide (Vickery 1993), we doubt that our results are applicable to other grassland systems. In Illinois, Herkert (1991b) rarely encountered Bobolinks on plots smaller than 20 ha and considered 10 to 30 ha to be the species' minimum area requirement. The minimum area for Eastern Meadowlarks in Illinois was smaller than 10 ha; its incidence was approximately 45 percent in the 0- to 20-ha class but increased to 100 percent in the 20- to 100-ha class (Herkert 1991b). Samson (1980a) found Eastern Meadowlarks on 100 percent of the sites ($n = 4$) smaller than 10 ha.

The general pattern for the six species found to be area-dependent in this study was similar to patterns found in Illinois (Herkert 1991b) and Missouri (Samson 1980a), despite the fact that recent herbicide use had changed the vegetation of some of the grassland-barrens in Maine (Yarborough and Bhowmik 1988). The consistency of these results suggests that these grassland/prairie species have similar area requirements across much of their range and that this is an important and consistent feature of each species' breeding ecology.

Species with Negative Area Effects

The incidence functions of three early-successional species—Song Sparrow, Common Yellowthroat, and Brown Thrasher—decreased with increasing size of the barren (Fig. 3). This decline was not surprising because the larger grassland-barren sites were more intensively managed for commercial blueberry production, and therefore the relative proportion of shrubby habitat diminished with increasing area (P. D. Vickery unpubl. data). Also, the distance between shrub clumps and forest edge was likely to be greater on the larger sites, and thus these microsites were probably less attractive to edge species (Ehrlich et al. 1988).

Species Demonstrating No Area Effects

Although multiple regression analysis indicated a positive association with area for Field Sparrows (Table 2), in logistic regression analysis incidence was never greater than 10 percent. In contrast, Field Sparrow incidence in Illinois was more than 80 percent on sites 20 to

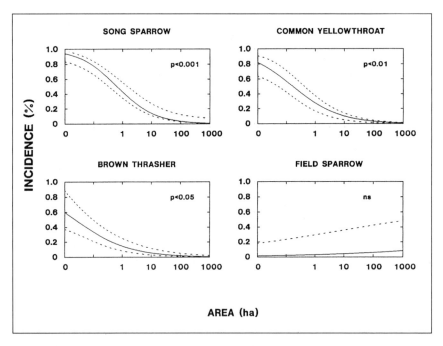

Figure 3. Incidence of three species of grassland birds was negatively correlated with increasing area of site. One species (Field Sparrow) showed no relationship with area. Dashed lines indicate 95 percent confidence intervals for the predicted probabilities.

100 ha in size (Herkert 1991b). As with Bobolinks and Eastern Meadowlarks, intensive management for blueberry production, which sharply reduced tall shrub and tree cover, made many of these sites unsuitable for Field Sparrows.

Effect of Area on Species Richness

The relationship between area and number of species per circular plot revealed higher species richness in the small size classes of 2 to 4 and 4 to 8 ha and in the largest size class of more than 64 ha ($r_s = 0.088$, not significant; Fig. 4). This pattern resulted from the preponderance of edge species such as Common Yellowthroats and Song Sparrows in the smaller size classes and the increase in area-dependent grassland species, notably Upland Sandpipers and Grasshopper Sparrows, in size classes of more than 32 ha. The total number of individuals per circular

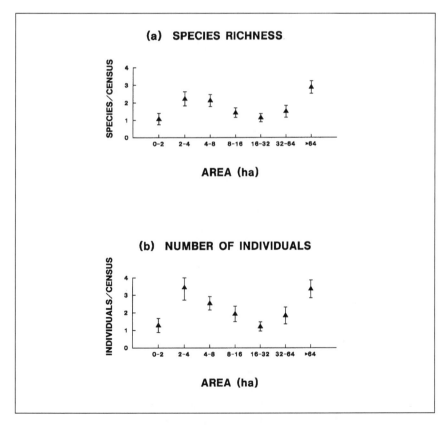

Figure 4. Relationship of (a) species richness per census plot ($X \pm 1$ SE) to area and (b) the number of individuals per census plot ($X \pm 1$ SE) to area on 90 grassland-barrens in coastal Maine, 1989–1990.

plot in each size class displayed a similar pattern to, and was strongly correlated with, total number of species ($r_s = 0.947$, $p < 0.001$; Fig. 4).

Raptors

Single Northern Harriers were observed on two occasions, both times beyond the perimeter of a circular plot on a site larger than 100 ha. Northern Harriers and Short-eared Owls have previously been known to breed, or occur in summer, on some of Maine's larger grassland-barrens (Adamus 1988, N. Famous pers. comm.). The area requirement for both these species is probably much greater than 100 ha (Gibbs et al. 1991, Serrentino 1992, Tate 1992). Their absence on these study sites can be partially attributed to the fact that adequate shrub cover necessary for nesting (Serrentino 1992, Tate 1992) had been reduced or eliminated by frequent burning and persistent herbicide use. These practices also substantially reduce small mammal populations, further reducing the suitability of the habitat for hawks and owls (Vickery et al. 1992). Also, daytime censusing was not appropriate for accurate assessment of Short-eared Owl incidence and abundance.

Regional Size Distribution Patterns of Grassland-barrens versus Hayfields-pastures

The size distribution of 100 randomly selected grassland-barren sites was different from the size distribution of 100 randomly selected hayfields-pastures within the same study area ($G = 21.74$, df = 6, $p < 0.001$; Fig. 5). Compared to hayfields-pastures, the proportion of grassland-barren sites increased with increasing size classes. This difference points out two important features: (1) there were few large hayfields and pastures in the study area (only 1 percent were larger than 64 ha), and (2) grassland-barrens provided most of the large grassland sites in this study area (81 percent were larger than 32 ha).

Conservation Implications and Recommendations

If conservation measures to protect threatened grassland birds are to be successful, they need to recognize and incorporate area-dependency and minimum viable population requirements (Shaffer 1981, Pimm et al. 1988) into management strategies (Samson 1980a, b). It is noteworthy that three of the species in this study that showed a strong positive correlation with area—Upland Sandpiper, Grasshopper Sparrow, and Vesper Sparrow—have been identified as regionally jeopardized in the New England–New York area (Vickery 1992). Northern Harrier and Short-eared Owl, both area-dependent grassland/heathland

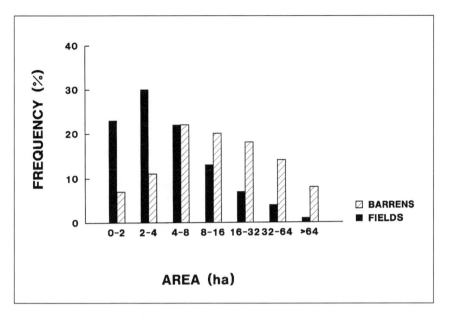

Figure 5. Frequency distribution of 100 randomly selected hayfields-pastures compared to 100 randomly selected grassland-barrens in coastal Maine, 1989–1990.

breeders, are also regionally jeopardized (Vickery 1992). Despite the differences in soil and current vegetation structure between grassland-barrens and native prairie, results from this and other studies have demonstrated that habitat size is a critical criterion for most grassland birds in these grassland habitats. Sites need to be larger than 50 ha, and preferably at least 200 ha, if they are to support a diverse grassland scolopacine and passerine fauna, and much larger if they are to support Northern Harriers or Short-eared Owls. Reproductive success is also likely to be greater in larger sites (Andren and Angelstam 1988, Johnson and Temple 1990). However, few sites exist in Maine and the Northeast that can support a full array of grassland birds. Therefore, the large grassland-barrens found in Maine have considerable regional significance. Relatively small grassland sites, comprising 5 to 10 ha, probably have a beneficial effect as secondary breeding sites for Vesper and Savannah sparrows (Brown and Kodric-Brown 1977), but they are unlikely to support raptors, Upland Sandpipers, or Grasshopper Sparrows, all of which prefer sites larger than 200 ha.

Acknowledgments

Various landowners in south, central, and eastern Maine generously permitted access to the sites in this study. Officials in the U.S. Soil Conservation Service, notably R. Dupuis, L. Hodgman, and C. Mitchell, provided aerial photographs, planimeters, and technical advice. The manuscript benefitted from discussions with W. E. Glanz, W. Halteman, J. Herkert, C. Smith, J. Wells, and A. S. White. Thoughtful reviews were provided by R. Askins, J. Gibbs, W. E. Glanz, G. L. Jacobson, Jr., R. O'Connor, B. Vickery, A. S. White, N. T. Wheelwright, and two anonymous reviewers. The Nongame Project of the Maine Department of Inland Fisheries and Wildlife provided financial support for the 1989 field-season research, which was conducted by Shawn Pierce. Critical logistical support was provided by D. Coonradt and B. Vickery.

Literature Cited

Adamus, P. R. 1988. Atlas of breeding birds in Maine, 1978-1983. Me. Dept. Inland Fish. Wildl., Augusta.

Ailes, I. W. 1980. Breeding biology and habitat use of the Upland Sandpiper in central Wisconsin. Passenger Pigeon 42: 53-63.

Andren, H., and P. Angelstam. 1988. Elevated predation rates as an edge effect in habitat islands. Ecology 69: 544-547.

Askins, R. A. 1993. Population trends in grassland, shrubland, and forest birds in eastern North America. Current Ornithol. 11: 1-34.

Bollinger, E. K., and T. A. Gavin. 1992. Eastern Bobolink populations: ecology and conservation in an agricultural landscape. Pp. 497-506 *in* Ecology and conservation of neotropical migrant landbirds (J. M. Hagan III and D. W. Johnston, eds.). Smithsonian Inst. Press, Washington, D.C.

Bond, R. R. 1957. Ecological distribution of breeding birds in the upland forests of southern Wisconsin. Ecol. Monogr. 27: 351-384.

Brown, J. H., and A. Kodric-Brown. 1977. Turnover rates in insular biogeography: effect of immigration on extinction. Ecology 58: 445-449.

Buhnerkempe, J. E., and R. L. Westemeier. 1988. Breeding biology and habitat of Upland Sandpipers on prairie-chicken sanctuaries in Illinois. Trans. Ill. Acad. Sci. 81: 153-162.

Canfield, R. 1941. Application of the line interception method in sampling range vegetation. J. Forestry 39: 388-394.

Diamond, J. M. 1975. The island dilemma: lessons of modern biogeographic studies for the design of nature reserves. Biol. Conserv. 7: 129-146.

Edwards, D. K., G. L. Dorsey, and J. A. Crawford. 1981. A comparison of three avian census methods. Stud. Avian Biol. 6: 170-176.

Ehrlich, P. R., D. S. Dobkin, and D. Whete. 1988. The birder's handbook. Simon and Schuster, New York.

Ferguson, R. H., and N. P. Kingsley. 1972. The timber resources of Maine. Research Bull. NE-26, U.S. Forest Serv., Northeast Forest Exper. Stn., Broomall, PA.

Fischer, R. A. 1954. Calendar of the letters of Alexander Baring, 1795-1801. Manuscript Div., Library of Congress, Washington, D.C.

Gibbs, J. P., J. R. Longcore, D. G. McAuley, and J. K. Ringelman. 1991. Use of wetland habitats by selected nongame water birds in Maine. Fish Wildl. Res. 9., U.S. Fish Wildl. Serv., Washington, D.C.

Haila, Y. 1986. North European land birds in forest fragments: evidence for area effects? Pp. 315-319 *in* Wildlife 2000: modeling habitat relationships of terrestrial vertebrates (J. Verner, M. Morrison, and C. J. Ralph, eds.). Univ. of Wisconsin Press, Madison.

Herkert, J. R. 1991a. Prairie birds of Illinois: population response to two centuries of habitat change. Ill. Nat. Hist. Surv. Bull. 34: 393-399.

Herkert, J. R. 1991b. An ecological study of the breeding birds of grassland habitats within Illinois. Ph.D. diss., Univ. of Illinois, Urbana-Champaign.

Johnson, R. G., and S. A. Temple. 1986. Assessing habitat quality for birds nesting in fragmented tallgrass prairies. Pp. 245-249 *in* Wildlife 2000: modeling habitat relationships of terrestrial vertebrates (J. Verner, M. L. Morrison, and C. J. Ralph, eds.). Univ. of Wisconsin Press, Madison.

Johnson, R. G., and S. A. Temple. 1990. Nest predation and brood parasitism of tallgrass prairie birds. J. Wildl. Manage. 54: 106-111.

Lanyon, W. E. 1957. The comparitive biology of the meadowlarks (*Sturnella*) in Wisconsin. Nuttall Ornithol. Club no. 1, Cambridge, MA.

O'Connor, R. J. 1981. Habitat correlates of bird distribution in British census plots. Stud. Avian Biol. 6: 533-537.

Pimm, S. L., H. L. Jones, and J. Diamond. 1988. On the risk of extinction. Am. Nat. 132: 757-785.

Robbins, C. S. 1979. Effect of forest fragmentation on bird populations. Pp. 198-212 *in* Management of north central and northeastern forests for nongame birds (R. M. DeGraaf and K. E. Evans, eds.). Gen. Tech. Rep. NC-51, U.S. Forest Serv., St. Paul, MN.

Robbins, C. S., D. Bystrak, and P. H. Geissler. 1986. The breeding bird survey: its first 15 years, 1965-1979. U.S. Fish Wildl. Serv. Res. Publ. 157.

Robbins, C. S., D. K. Dawson, and B. A. Dowell. 1989. Habitat area requirements of breeding forest birds of the middle Atlantic states. Wildl. Monogr. 103: 1-34.

Rudnicky, T. C., and M. L. Hunter, Jr. 1993. Reversing the fragmentation perspective: effects of clearcut size on bird species richness in Maine. J. Ecol. Appl. 3: 357-366.

Samson, F. B. 1980a. Island biogeography and the conservation of prairie birds. Proc. N. Am. Prairie Conf. 7: 293-305.

Samson, F. B. 1980b. Island biogeography and the conservation of nongame birds. Trans. N. Am. Wildl. Nat. Res. Conf. 45: 245-251.

Serrentino, P. 1992. Northern Harrier (*Circus cyaneus*). Pp. 89-117 *in* Migratory nongame birds of management concern in the northeast (K. J. Schneider and D. M. Pence, eds.). U.S. Fish Wildl. Serv., Newton Corner, MA.

Shafer, C. L. 1990. Nature reserves. Smithsonian Inst. Press, Washington, D.C.

Shaffer, M. L. 1981. Minimum population sizes for species conservation. Bioscience 31: 131-134.

Smith, D. J., and C. R. Smith. 1992. Henslow's Sparrow and Grasshopper Sparrow: a comparison of habitat use in Finger Lakes National Forest, New York. Bird Observer 20(4): 187-194.

Sokal, R. R., and F. J. Rohlf. 1981. Biometry. 2d ed. W. H. Freeman, New York.

Tate, G. R. 1992. Short-eared Owl (*Asio flammeus*). Pp. 171-189 *in* Migratory nongame birds of management concern in the northeast (K. J. Schneider and D. M. Pence, eds). U.S. Fish Wildl. Serv., Newton Corner, MA.

U.S. Department of Agriculture. 1936-1991. Agricultural statistics. U.S. Gov. Printing Office, Washington, D.C.

Verner, J. 1985. Assessment of counting techniques. Pp. 247-302 *in* Current ornithology, vol. 2 (R. F. Johnston, ed.). Plenum Press, New York.

Vickery, P. D. 1990. Report on grassland habitats in relation to the presence of Grasshopper Sparrows (*Ammodramus savannarum*) and other rare vertebrates in Maine. Maine Natural Heritage Prog., Office Comprehensive Planning, Augusta.

Vickery, P. D. 1992. A regional analysis of endangered, threatened, and special-concern birds in the northeastern United States. Trans. N. E. Sect. Wildl. Soc. 48: 1-10.

Vickery, P. D. 1993. Habitat selection of grassland birds in Maine. Ph.D. diss., Univ. of Maine, Orono.

Vickery, P. D., M. L. Hunter, Jr., and J. V. Wells. 1992. Evidence of incidental nest predation and its effects on nests of threatened grassland birds. Oikos 63: 281-288.

Wilkinson, L. 1990. Systat: the system for statistics. Systat, Inc., Evanston, IL.

Winne, J. C. 1988. History of vegetation and fire on the Pineo Ridge blueberry barrens in Washington County, Maine. Master's thesis, Univ. of Maine, Orono.

Yarborough, D. E., and P. C. Bhowmik. 1988. Effect of Hexazinone on weed populations and on lowbush blueberries in Maine. Acta Horticult. 241: 344-349.

Zar, J. H. 1984. Biostatistical analysis. Prentice-Hall, Englewood Cliffs, NJ.

Peter D. Vickery: Center for Biological Conservation, Massachusetts Audubon Society, Lincoln, MA 01773, and Department of Forestry and Wildlife Conservation, University of Massachusetts, Amherst, MA 01003.

Malcolm L. Hunter, Jr.: Wildlife Department, University of Maine, Orono, ME 04469.

Scott M. Melvin: Endangered and Nongame Wildlife Project, Maine Department of Inland Fisheries and Wildlife, Bangor, ME 04401 (present address: Natural Heritage and Endangered Species Program, Division of Fisheries and Wildlife, Westborough, MA 01581).

Population Viability Analysis for Maine Grasshopper Sparrows

Jeffrey V. Wells

Abstract

The Grasshopper Sparrow (*Ammodramus savannarum*) is among a suite of grassland birds showing regional and continentwide declines. In Maine the species is state endangered, yet no comprehensive management plan has been developed. Using demographic data obtained from recent censuses and a banding study of Maine Grasshopper Sparrow populations, I used a stochastic simulation model to compute probabilities of extinction under various conditions. Given current demographic parameters, the model estimated that there was a 50 percent probability that Maine Grasshopper Sparrows will become extinct within 50 years. Simulated loss or degradation of habitat at one of the smaller populations increased extinction risk to 73 to 79 percent. Simulated optimum management of the habitat of all Maine Grasshopper Sparrow populations decreased extinction risk to 22 percent. Sensitivity analyses showed that decreasing the variance in population growth rates, particularly of the state's largest population, would greatly decrease the probability of extinction. A simple model demonstrates how managers can decrease variance in population growth rate by optimizing high-quality habitat.

Introduction

Population Viability Analysis (Shaffer 1991), or risk analysis (Burgman et al. 1993), has become an increasingly important tool to assist biological managers in decision-making. The technique allows the quantification of extinction risks under various management options. This permits objective consideration of management plans rather than subjective assessments, which are difficult to defend and usually underestimate risks (Zeckhauser and Viscusi 1990, Burgman et al. 1993). Furthermore, the model development process of risk analysis helps identify important demographic parameters and focuses attention on data needs (Boyce 1992). Where there are limited financial resources, such information is vital for prioritizing research and management objectives. Risk analysis does require common sense, however, and caution in interpretation. For example, a risk analysis that led to the recommendation that a captive population of Puerto Rican Parrots (*Amazona vittata*) be subdivided among several facilities did not consider the potential for contraction of communicable diseases in mainland zoological institutions, a factor that may increase mortality significantly (Wilson et al. 1994).

Risk analyses have been carried out for numerous animal and plant species, including butterflies (Murphy et al. 1990), perennial plants (Menges 1990), elephants (Armbruster and Lande 1993), and woodpeckers (Haig et al. 1993). Perhaps the best-known species for which risk analyses have been done is the Northern Spotted Owl (*Strix occidentalis caurina*); various management plans have been proposed and debated for this species (Lande 1988, Thomas et al. 1990, Anderson and Burnham 1992, Lamberson et al. 1992, McKelvey et al. 1992, Murphy and Noon 1992, Harrison et al. 1993).

The Grasshopper Sparrow is among a suite of grassland bird species showing regional and continentwide declines (Vickery 1992, Askins 1993, Johnson and Schwartz 1993). In Maine the species is state endangered and is known to breed at only four sites. However, a management plan has not been developed (C. Todd. pers. comm.). Risk analyses have not been carried out for the Grasshopper Sparrow or any other declining grassland bird species. Therefore, I carried out a risk analysis to determine the probability of extinction of Maine Grasshopper Sparrows under a variety of possible management options and to determine which factors might contribute most to extinction risk.

Methods

I used RAMAS-space, a metapopulation simulation model (Akcakaya and Ferson 1990), which has been widely used and tested (Akcakaya and Ginzburg 1991, Akcakaya 1992, Buckley and Downer 1992, Conroy 1992, Ferson and Akcakaya 1992, Lamberson 1992, Burgman et al. 1993, Gibbs 1993, LaHaye et al. 1994, Akcakaya et al. 1995, Lindenmeyer et al. 1995), to assess extinction risks. I used a logistic model of the form N(t+1) = N(t) * $e^{r * (K-N(t)/K)}$ (Akcakaya and Ferson 1990), with population growth rate (Lambda = e^r), carrying capacity, and survivorship set independently for each population. Populations were then linked by rates of dispersal; these rates were set independently for each pair of populations. Correlations between population growth rates were set independently for each pair of populations to account for the likelihood that sites near each other would experience similar environmental conditions. The program also allowed the incorporation of environmental and demographic stochasticity.

Maine Grasshopper Sparrows are distributed in four spatially disjunct populations (as defined in Wells and Richmond 1995). Three of the populations are within 10 kilometers (km) of one another, whereas the fourth (Brunswick, Cumberland Co.) is 120 km northeast. The Brunswick population is very small (only one singing male present in 1993 and 1994) and is probably independent of the other populations. For these reasons I did not consider the Brunswick population in this analysis. Regular population surveys have taken place at all four populations for 7 years (1988–1994) and at some for 11 years (1984–1994; Fig. 1).

Habitat modification as a result of herbicide application from 1984 to 1987 at Kennebunk caused a constant decrease in abundance over a four- to six-year period so that only post-habitat-loss population surveys could be used to estimate demographic parameters for that site (there were no preherbicide abundance estimates). Herbicide application at Wells occurred in 1989, apparently causing Grasshopper Sparrows to abandon the site. I therefore excluded that year from the estimates of demographic parameters for Wells. Finite rate of increase (Lambda) for each of the three populations was calculated as the mean proportional change in the number of singing males over the survey period, and standard deviation in Lambda was calculated for each population from the same data (Table 1). Initial abundances were taken as the 1994 estimated abundances.

Probability of survival was taken as the average return rate of male Grasshopper Sparrows individually color-banded at Kennebunk over a four-year period (Wells 1993). Correlations among populations were calculated as the correlation between annual proportional change in abundance (number of singing males) for each pair of populations (Table 2). Few juvenile Grasshopper Sparrows were banded at Kennebunk, so there is no precise estimate of the amount of dispersal between populations. However, it is clear that there is not a great deal

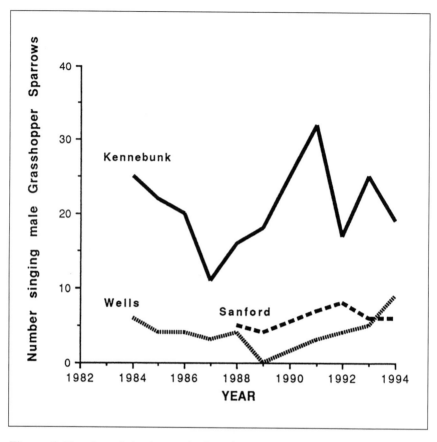

Figure 1. Number of singing male Grasshopper Sparrows at the three southern Maine breeding sites since surveys began at the respective locations. Demographic parameters were estimated using data from 1988 to 1994 at Kennebunk and from 1984 to 1988 and 1991 to 1994 at Wells because of herbicide-induced population changes at the sites in the years omitted (1984–1987 at Kennebunk, 1989–1990 at Wells). Surveys did not begin at Sanford until 1988; demographic parameters for the population at this site were calculated using data from 1988 to 1994.

of movement of individuals between populations, because of 72 individuals banded at Kennebunk (including 19 juveniles, 2 of which returned and bred within 500 meters [m] of their parents' territory), none were sighted elsewhere despite thorough searches at Wells and Sanford (all singing males present at both sites were checked for bands from 1991 through 1994). I therefore used an arbitrary estimate of 1 percent dispersal from Kennebunk to the other two populations each year, though I also ran sensitivity analyses to assess the effect of

Table 1. Current best-estimate values for demographic parameters used in model to estimate extinction risks for Maine Grasshopper Sparrows. These values are based on data collected as late as 1994. See text for details.

Site	Lambda	SD	Surv	K	Correlations		
					Kenn	Wells	San
Kennebunk	1.0917	0.3725	0.345	150	1.000	-0.409	-0.992
Wells	1.1614	0.3893	0.345	80	-0.409	1.000	0.292
Sanford	0.9225	0.1808	0.345	40	-0.992	0.292	1.000

Lambda = population growth rate
SD = standard deviation in population growth rate
Surv = survivorship
K = carrying capacity
Kenn = Kennebunk
Wells = Wells
San = Sanford

Table 2. Demographic parameters used in sensitivity analyses to model extinction risks for Maine Grasshopper Sparrows. These values are based on average values from data collected as late as 1994. See text for details.

Site	Lambda	SD	Surv	K	Correlations		
					Kenn	Wells	San
Kennebunk	1.0585	0.3142	0.345	150	1.000	-0.362	-0.362
Wells	1.0585	0.3142	0.345	80	-0.362	1.000	-0.362
Sanford	1.0585	0.3142	0.345	40	-0.362	-0.362	1.000

Lambda = population growth rate
SD = standard deviation in population growth rate
Surv = survivorship
K = carrying capacity
Kenn = Kennebunk
Wells = Wells
San = Sanford

different levels of dispersal (see below). Because of the small sizes of the Wells and Sanford populations, I assumed there would be no dispersal from these populations to each other or to Kennebunk.

A logistic model was used to incorporate possible density-dependent effects at very high densities to effectively place an upper bound on population size. I used a carrying capacity for each population (Table 2) that would result in a density higher than typically found anywhere within the range of Grasshopper Sparrow (high densities range from 27 to 40 males per 100 hectares [ha; Allaire 1980, Johnson and Schwartz 1993], whereas carrying capacities for my model were set at values that would result in densities of approximately 60 males per 100 ha). This was done to alleviate the influence of density dependence except at very high abundances because density dependence is unlikely to be a major factor in these populations in the short term. Thus it is likely that the estimated extinction risks will be conservative (i.e., extinction risk tends to be higher if density dependence is not operating [Ginzburg et al. 1990]).

Stochastic variation in population growth rates was incorporated by randomly selecting each simulated population's annual growth rate from a lognormal distribution with the designated mean and standard deviation. Annual survivorship was drawn from a binomial distribution with the specified mean. For each set of demographic parameters, I ran 500 simulations following the recommendations of Harris et al. (1987). All simulations were run for a 50-year period.

All models are limited by the assumptions they incorporate into the model. Therefore, such limitations need to be known to evaluate model results. In the case of the model used in this analysis, two of the most important limitations were the lack of age structure and the lack of consideration of genetic effects. It is unclear how the addition of age structure would have affected the model results because the finite rate of increase and the probability of survival were effectively averaged over all age classes in my model. The addition of genetic considerations to the model would likely (but not necessarily) have increased extinction risk because of the loss of heterozygosity that often accompanies small populations.

For sensitivity analyses, I used as base values the average of the three populations' growth rates and standard deviations. Correlations were taken as the average of correlations in proportional change in abundance. For sensitivity analyses, then, all three populations had the same average and standard deviation of growth rate, and correlations among all populations were the same (Table 2). Survivorship was set at the average of 0.345 (Table 2), and initial abundances were 1994 estimated abundances. I then varied each parameter while holding all other parameters constant to test how variation in each parameter affected extinction risk.

To test the effects of various management options on extinction risk for Maine Grasshopper Sparrows, I used the estimated parameters from Table 1 and varied them in the following ways:

1. decreased growth rate at the Wells population from 1.1614 to 0.9225 (Sanford's growth rate) to simulate the effect of habitat degradation on population growth rate;

2. caused the carrying capacity of the Wells population to decline by 5 breeding pairs per year (from 80) to simulate slow annual habitat loss;

3. decreased standard deviation in population growth rate for the Kennebunk population from 0.3725 to 0.1808 (Sanford's standard deviation) to simulate a habitat-management scheme that optimized good-quality habitat over the long term;

4. changed the population growth rates at all populations to the Wells average (1.1614) to simulate the effect of managing those sites specifically for Grasshopper Sparrows; and

5. added immigrants from a fourth source population at average rates of 1, 3, 6, and 9 individuals per year. The source population itself was not included in the calculation of extinction risk because what was of interest was the effect immigrants from a source population would have on the extinction risk of the Maine metapopulation.

To demonstrate how different habitat-management schemes (in this case, different burn rotations) could affect variance in population growth rate, I developed a simple model in which three areas are burned once every four years. In the model there were three habitat states: excellent, good, and poor quality. Excellent-quality habitat was grassland burned two years ago; good-quality habitat was grassland burned one year ago; and poor-quality habitat was grassland burned that year. These habitat-quality assessments have been shown to hold true for Grasshopper Sparrow and other grassland species after prescribed burns in Maine (Vickery 1993). Each pair of birds in excellent-quality habitat fledged an average of 3 young per year, each pair in good-quality habitat fledged an average of 1.5 young per year, and those in poor-quality habitat fledged an average of only 0.5 young per year.

I contrasted three burn rotations. In one, the entire area was burned every four years. In another, 50 percent of the area was burned one year, 50 percent the next year, then no burning took place for two years, after which the burn schedule was repeated. In the third rotation, 25 percent of the area was burned each year. Each area began with 20 pairs of birds distributed uniformly over the area. I then computed population trajectories for each burn-rotation scenario over six seasons based on the number of pairs within each habitat state. For example, in the rotation where 25 percent of the area was burned each year, 15 pairs were in excellent-quality habitat the first year with an average productivity of 3.0, and 5 pairs were in poor-quality habitat with an average productivity of 0.5. These 20 pairs produced 47.5 young in the first season. From these population trajectories, I calculated average growth rates and variance in growth rate for each burn-rotation scheme.

Results

I first computed the probability of extinction within 50 years for each of the three southern Maine Grasshopper Sparrow populations independently. That is, I considered what each population's extinction risk would be if there were no movement of individuals among populations and if there were no environmental correlation among populations. Under these conditions, the Sanford population would face almost certain extinction (100 percent) within 50 years, and the Wells and Kennebunk populations would face high probabilities of extinction (76 and 71 percent, respectively). To compute the probability of extinction of the three Maine Grasshopper Sparrow populations (assuming independence), the extinction probabilities for each of the three sites had to be multiplied. This resulted in a 50-year extinction risk for the three independent populations of 52 percent.

The three populations are not independent of each other, however. Individuals must at least occasionally move among these sites, and the populations must have some degree of environmental correlation. Thus, this set of three populations can be considered a metapopulation (Wells and Richmond 1995). Based on simulations using the values in Table 1, the current estimate of extinction risk for the Maine Grasshopper Sparrow metapopulation within the next 50 years is 50 percent.

Effects of Different Management Options on Metapopulation Extinction Risk

A decrease in the standard deviation in growth rate at the Kennebunk population from 0.3725 to 0.1808 decreased the metapopulation extinction risk to 35 percent. Adding just a single immigrant per year decreased extinction risk to 20 percent; adding six immigrants per year dropped extinction risk to zero (Fig. 2). Decreasing the population growth rate at Wells from 1.1614 to 0.9225 increased metapopulation extinction risk from 50 to 73 percent. A decrease in carrying capacity of five breeding pairs per year at Wells to simulate gradual habitat loss increased metapopulation extinction risk to 79 percent. Increasing the population growth rates of all three populations to the value at Wells (1.1614) decreased extinction risk to 22 percent.

Sensitivity Analyses

Varying correlation values from 0.0 to 1.0 resulted in extinction probabilities ranging from 58 to 68 percent (Fig. 3A). Simultaneously varying the dispersal rate among all three populations from 0.01 to 0.4 resulted in extinction probabilities ranging from 32 to 47 percent (Fig. 3B). Varying each population's rate of dispersal to the other two

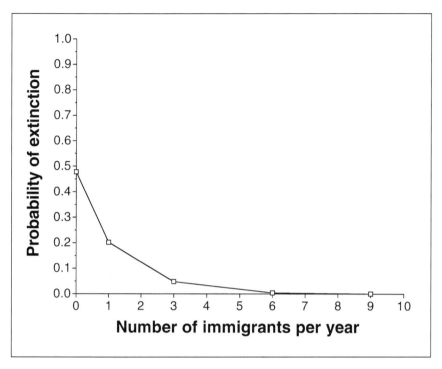

Figure 2. *Probability of extinction for the Maine Grasshopper Sparrow metapopulation with varying annual numbers of immigrants from an outside source population.*

populations from 0.05 to 0.45 showed that the rate of dispersal from Kennebunk to the other populations had the greatest effect on extinction risk (Fig. 3C), though the effect was small. Simultaneously varying standard deviations in growth rate from 0.0 to 1.0 for all three populations resulted in extinction probabilities ranging from 29 to 99 percent (Fig. 3D). By varying the standard deviation in growth rate for each population, it became clear that variance of the growth rate for the Kennebunk population had the greatest effect on extinction risk (Fig. 3E). Not unexpectedly, varying population growth rate had a large effect on extinction risk. Varying growth rate from 0.8 to 1.3 caused extinction risk to range from 100 to 4 percent (Fig. 3F).

Burn-rotation Model

The simple burn-rotation model showed the greatest increase in abundance when 25 percent of the area was burned each year (Fig. 4). Burning 25 percent of the area each year resulted in the lowest variance (0.2874), whereas burning the entire area every four years

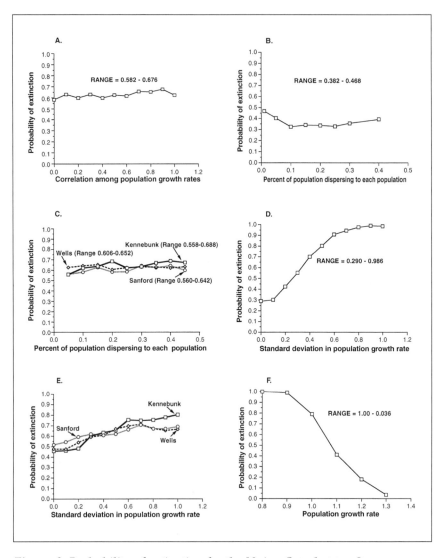

Figure 3. Probability of extinction for the Maine Grasshopper Sparrow metapopulation under a variety of conditions. Values for each specific analysis range over (A) correlations in population growth rates, (B) among-population dispersal rates, (C) population dispersal rates, (D) standard deviation in population growth rate, for which the standard deviations were the same for all populations, (E) standard deviation in population growth rate, for which the standard deviation for each population was varied while leaving the other populations at fixed values, and (F) differing population growth rates, for which all populations were given the same population growth rate. See Table 2 for values of all parameters.

resulted in the highest variance (0.7463). Burning 25 percent of the habitat each year maximized the amount of best-quality habitat over the long term, which in turn decreased the variance in population growth rate and therefore the risk of extinction for the population.

Discussion

The best estimate of the risk of extinction for Maine Grasshopper Sparrows within the next 50 years is 50 percent. This estimate, however, is based on relatively short-term demographic data. To illustrate the limitations of short-term data, the probability of extinction for the Maine Grasshopper Sparrow metapopulation based on pre-1994 data was 39 percent compared to the 50 percent probability of extinction obtained with the inclusion of 1994 data. These differences were largely the result of a 24 percent population decline at Kennebunk and an 80 percent population increase at Wells. Long-term estimates of

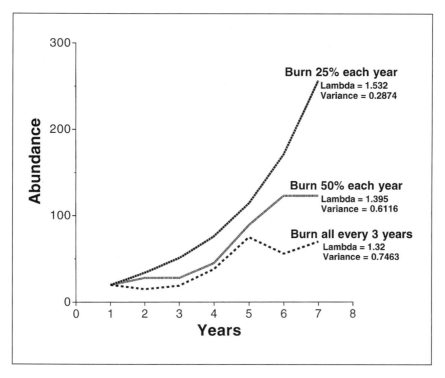

Figure 4. *Population trajectories for model populations under each of the three possible burn-rotation schemes outlined in the text. Maximizing the amount of available optimum habitat decreases variance in population growth rates, which in turn decreases extinction risk.*

population size for all three populations would increase confidence in estimates of average population growth rate, its variance, and correlation between population growth rates, which would therefore increase confidence in assessment of extinction risk. Clearly, continued commitment to surveys of population size is important for those interested in assessing extinction risk for Maine Grasshopper Sparrows.

Determining the current extinction risk for Maine Grasshopper Sparrows with short-term demographic data is of limited utility. However, examining the effect of various management options on extinction risk, even with short-term data, is more meaningful. With this approach, one can compare the relative extinction risks among different management alternatives without being overly concerned with the exact extinction-risk value. Therefore, I carried out simulations to explore alternative strategies for ensuring the existence of breeding populations of Grasshopper Sparrows in Maine. The effect of a management strategy to optimize high-quality habitat over the long term at Kennebunk was simulated by decreasing variance in population growth rate from 0.3725 to 0.1808. This decreased extinction risk for the entire Maine Grasshopper Sparrow metapopulation from 50 to 35 percent. A management strategy at Kennebunk that optimized high-quality habitat over the long term would decrease extinction risk for the entire metapopulation.

It is difficult to assess the true level of immigration into the Maine metapopulation. However, it is clear that even a very small number of immigrants can have a great effect on extinction risk. It is important, therefore, for Maine wildlife managers to encourage and cooperate with wildlife managers from other New England states (notably Massachusetts) to ensure that potential source populations remain healthy. Management plans for species that occur in adjoining states should be developed cooperatively with officials from the various states. Interstate cooperation is especially relevant to grassland bird species in the northeastern United States; many species are state listed in this region (Vickery 1992) and are exhibiting regional declines (Askins 1993).

Management plans for species that occur on lands managed by more than one agency need similar coordination. Grassland bird species often occur on lands owned or managed by groups as varied as the U.S. Air Force, The Nature Conservancy, and private landowners, so that effective regional management requires a high degree of cooperation. In Massachusetts, such cooperative efforts have had great success; a cooperative management plan at Westover Air Force Base in Chicopee, for example, tripled the number of Grasshopper Sparrows at the site over a seven-year period (Melvin 1994, Jones and Vickery 1995).

One of the sites where Grasshopper Sparrows occur in Maine (Wells) is privately owned with no management agreement in place. Simulating the effect of habitat degradation at the site increased extinction risk for the entire metapopulation from 50 to 73 percent, whereas simulating the effect of habitat loss increased extinction risk from 50 to 79 percent. Concerted habitat management at all three sites, however, would potentially decrease extinction risk from 50 to 22 percent. There is little doubt that securing a management agreement at Wells and Sanford would be beneficial to Maine Grasshopper Sparrows.

Sensitivity analyses showed that correlation between population growth rates had a relatively small effect on extinction risk (a 10,000 percent increase in correlation resulted in a 10 percent increase in extinction risk), apparently because of the small size of the Wells and Sanford populations. However, increased correlation did cause an increase in extinction risk. An increase in extinction risk with increasing environmental correlation (a high degree of environmental correlation means that populations experience similar environmental conditions at the same time) has been shown to be a general result in metapopulation models (Goodman 1987, Burgman et al. 1993). In many current grassland habitats, periodic mowing and/or burning is required to maintain the desired plant and animal communities. The environmental correlation between nearby populations can be strongly affected by whether or not the habitat management rotations at the sites are the same. For example, if two nearby sites are burned in the same year, the correlation in population growth rates between the sites will increase. Although the effect may be slight, managers of grassland habitats that are near each other should coordinate habitat-management efforts to decrease environmental correlation between sites, reducing the risk of extinction.

Extinction risk decreased from 47 percent with a dispersal rate of 1 percent among populations to 32 percent with a dispersal rate of 10 percent among populations (Fig. 3B). An increase in dispersal rate beyond 10 percent had little effect on extinction risk. Better estimates of dispersal rates among populations would improve the predictive power of the model. However, the model is relatively insensitive to differences in dispersal rate, in contrast to the model's sensitivity to growth rate and variance in growth rate.

The two parameters to which the model was most sensitive were population growth rate and its variance. Variance in population growth rate had a significant effect on extinction risk (a 10,000 percent increase in standard deviation resulted in a 70 percent increase in extinction risk), particularly variance in population growth rate of the state's largest population (Kennebunk). Other modeling efforts have shown that variance in population growth rates is often the most important demographic factor determining extinction risk (Goodman 1987, Ruggiero et al. 1994). For many species, it may be quite difficult to find ways to manage for lower variance in population growth rates (Goodman 1987). However, the relatively frequent habitat management required of grasslands offers opportunities to affect variance in population growth rate.

Using a simple model, I showed that managers of grassland habitat can greatly affect the variance in population growth rate by employing habitat-management schemes that maximize high-quality habitat (habitat in which the population can achieve high densities, high reproductive success, and low mortality) over the long term. In the case of a 400-ha site in which the habitat required burning every four years, maximizing high-quality habitat over the long term would mean burning 100 ha per year. Although in theory this may seem relatively simple, in practice many factors can influence the implementation of habitat-management plans. For example, drought conditions in 1995 at Kennebunk (where a burn-rotation plan to maximize high-quality habitat is in place) limited the number of prescribed burns that could be carried out (P. Vickery pers. comm.).

Conclusions

1. Good estimates of the population growth rate and its variation are necessary to assess extinction risks confidently (Ruggiero et al. 1994). This requires long-term population census data. Short-term data may over- or underestimate extinction risks significantly, and therefore models based on such data should be interpreted with caution. There are, however, no clear guidelines for deciding whether or not a particular time series is long enough. If in a time series several years of population lows have been observed among a number of years of higher population sizes, one might feel some confidence in estimates of population growth rate since clearly the population has not been steadily increasing or decreasing. A reasonable rule of thumb is that data sets of fewer than 10 to 15 years should be considered short-term time-series data sets. Unfortunately, management decisions concerning endangered species are usually necessary before long-term data can be collected. Population Viability Analysis, even in the face of limited data, can elucidate which factors are most important in determining extinction risk (Boyce 1992, Burgman et al. 1993).

2. Maximizing optimum habitat over the long term increases the population growth rate and decreases its variation, thereby decreas-

ing extinction risk (Goodman 1987). Careful attention to burn-rotation (or other habitat-modification) schemes is important for minimizing extinction risks.

3. Increasing correlations among population growth rates increases metapopulation extinction risk. Burning or mowing rotations between nearby grassland habitats should be offset to decrease correlations and thereby decrease extinction risk.

4. Source populations are very important in decreasing metapopulation extinction risk. Therefore, managers of preserves under different ownership and political boundaries should cooperate to ensure the continued presence of desired species.

Acknowledgments

Fieldwork was supported by funds from the Benning Fund of the Cornell Laboratory of Ornithology, Cornell University, Eastern Bird Banding Association, International Council for Bird Preservation, Kathleen Anderson Award of Manomet Center for Conservation Sciences, Maine Chapter of The Nature Conservancy, Mellon Foundation, national office of The Nature Conservancy, New York Cooperative Fish and Wildlife Research Unit, Endangered/Threatened Species Unit of the Maine Department of Inland Fisheries and Wildlife, Sigma Xi Grants-in-Aid of Research, and Wilson Ornithological Society. A preliminary PVA analysis was funded through a contract with the Endangered/Threatened Species Unit of the Maine Department of Inland Fisheries and Wildlife. I especially thank C. Todd and A. Hutchinson for their support. Thanks to B. Askins, T. Gavin, M. Richmond, C. Smith, P. Vickery, A. C. Wells, and D. Winkler for providing constructive criticism on earlier drafts of the manuscript. This work was done in partial fulfillment of a Ph.D. degree from Cornell University.

Literature Cited

Akcakaya, H. R. 1992. Population viability analysis and risk assessment. Pp. 148–157 *in* Wildlife 2001: populations (D. R. McCullough and R. H. Barrett, eds.). Elsevier, London.

Akcakaya, H. R., and S. Ferson. 1990. RAMAS/space user manual. Applied Biomathematics, Setauket, NY.

Akcakaya, H. R., and L. R. Ginzburg. 1991. Ecological risk analysis for single and multiple populations. Pp. 73–87 *in* Species conservation: a population biological approach (A. Seitz and V. Loeschcke, eds.). Birkhauser Verlag, Basel.

Akcakaya, H. R., M. A. McCarty, and J. Pearce. 1995. Linking landscape data with population viability analysis: management options for the Helmeted Honeyeater. Biol. Conserv. 73: 169-176.

Allaire, P. N. 1980. Bird species on mined lands. Institute for Mining and Minerals Research, Kentucky Center for Energy Research Lab., Univ. of Kentucky, Lexington.

Anderson, D. R., and K. P. Burnham. 1992. Demographic analysis of Northern Spotted Owl populations. Recovery plan for the Northern Spotted Owl, appendix C. U.S. Fish Wildlife Serv., Portland, OR..

Armbruster, P., and R. Lande. 1993. A population viability analysis for African elephants (*Loxodonta africana*): how big should reserves be? Conserv. Biol. 7: 602-610.

Askins, R. A. 1993. Population trends in grassland, shrubland, and forest birds in eastern North America. Current Ornithol. 11: 1-34.

Boyce, M. S. 1992. Population viability analysis. Ann. Rev. Ecol. Syst. 23: 481-506.

Buckley, P. A., and R. Downer. 1992. Modelling dynamics for single species of seabirds. Pp. 563-585 *in* Wildlife 2001: populations (D. R. McCullough and R. H. Barrett, eds.). Elsevier, London.

Burgman, M. A., S. Ferson, and H. R. Akcakaya. 1993. Risk assessment in conservation biology. Chapman and Hall, London.

Conroy, M. J. 1992. Review of RAMAS space: spatially structured populations models for conservation biology. Quart. Rev. Biol. 68: 159-160.

Ferson, S., and H. R. Akcakaya. 1992. Quantitative software tools for conservation biology. Pp. 371-386 *in* Computer techniques in environmental studies IV (P. Zannetti, ed.). Computational Mechanics Publ., Southampton, and Elsevier Appl. Sci., London.

Gibbs, J. P. 1993. Importance of small wetlands for the persistence of local populations of wetland-associated animals. Wetlands 13: 25-31.

Ginzburg, L. R., S. Ferson, and H. R. Akcakaya. 1990. Reconstructibility of density dependence and the conservative assessment of extinction risks. Conserv. Biol. 4: 63-70.

Goodman, D. 1987. How do any species persist? Lessons for conservation biology. Conserv. Biol. 1: 59-62.

Haig, S. M., J. R. Belthoff, and D. H. Allen. 1993. Population viability analysis for a small population of Red-cockaded Woodpeckers and an evaluation of enhancement strategies. Conserv. Biol. 7: 289-301.

Harris, R. B., L. A. Maguire, and M. L. Shaffer. 1987. Sample sizes for minimum viable population estimation. Conserv. Biol. 1: 72-76.

Harrison, S., A. Stahl, and D. Doak. 1993. Spatial models and Spotted Owls: exploring some biological issues behind recent events. Conserv. Biol. 7: 951-953.

Johnson, D. H., and M. D. Schwartz. 1993. The conservation reserve program and grassland birds. Conserv. Biol. 7: 934-937.

Jones, A. L., and P. D. Vickery. 1995. Distribution and population status of grassland birds in Massachusetts. Bird Observer 23: 89-96.

LaHaye, W. S., R. J. Gutierrez, and H. R. Akcakaya. 1994. Spotted Owl metapopulation dynamics in southern California. J. Anim. Ecol. 63: 775-785.

Lamberson, R. 1992. Net notes: software review of RAMAS-age, stage, space. Nat. Res. Modeling 6: 99-102.

Lamberson, R. H., K. McKelvey, B. R. Noon, and C. Voss. 1992. A dynamic analysis of Northern Spotted Owl viability in a fragmented forest landscape. Conserv. Biol. 6: 505-512.

Lande, R. 1988. Demographic models of the Northern Spotted Owl (*Strix occidentalis caurina*). Oecologia 75: 601-607.

Lindenmeyer, D., M. Burgman, H. R. Akcakaya, R. Lacy, and H. Possingham. 1995. A review of the generic computer programs ALEX, RAMAS-space, and VORTEX for modelling the viability of wildlife metapopulations. Ecol. Modelling 82: 161-174.

McKelvey, K., B. R. Noon, and R. H. Lamberson. 1992. Conservation planning for species occupying fragmented landscapes: the case of the Northern Spotted Owl. Pp. 424-450 *in* Biotic interactions and global change (P. M. Kareiva, J. G. Kingsolver, and R. B. Huey, eds.). Sinauer, Sunderland, MA.

Melvin, S. M. 1994. Military bases provide habitat for rare grassland birds. Mass. Div. Fish. Wildl. Nat. Heritage News 4: 3.

Menges, E. S. 1990. Population viability analysis for an endangered plant. Conserv. Biol. 4: 52-62.

Murphy, D. D., K. E. Freas, and S. B. Weiss. 1990. An environmental-metapopulation approach to population viability analysis for a threatened invertebrate. Conserv. Biol. 4: 41-51.

Murphy, D. D., and B. R. Noon. 1992. Integrating scientific methods with habitat conservation planning: reserve design for Northern Spotted Owls. Ecol. Appl. 2: 3-17.

Ruggiero, L., G. D. Hayward, and J. R. Squires. 1994. Viability analysis in biological evaluations: concepts of population viability analysis, biological population, and ecological scale. Conserv. Biol. 8: 364-372.

Shaffer, M. 1991. Population viability analysis. Pp. 108-118 *in* Challenges in the conservation of biological resources (D. J. Decker, M. E. Krasny, G. R. Goff, C. R. Smith, and D. W. Gross, eds.). Westview Press, Boulder, CO.

Thomas, J. W., E. D. Forsman, J. B. Lint, E. C. Meslow, B. R. Noon, and J. Verner. 1990. A conservation strategy for the Northern Spotted Owl. Interagency Scientific Committee to Address the Conservation of the Northern Spotted Owl. U.S. Gov. Printing Office, Washington, D.C.

Vickery, P. D. 1992. A regional analysis of endangered, threatened, and special-concern birds in the northeastern United States. Trans. N. E. Sect. Wildl. Soc. 48: 1-10.

Vickery, P. D. 1993. Habitat selection of grassland birds in Maine. Ph.D. diss., Univ. of Maine, Orono.

Wells, J. V. 1993. Correlates of return rates of Grasshopper and Vesper sparrows at an isolated habitat island. Assoc. Field Ornithologists 1993 Annual Meeting Abstracts.

Wells, J. V., and M. E. Richmond. 1995. Populations, metapopulations, and species populations: what are they and who should care? Wildl. Soc. Bull. 23: 458-462.

Wilson, M. H., C. B. Kepler, N. F. R. Snyder, S. R. Derrickson, F. J. Dein, J. W. Wiley, J. M. Wunderle, Jr., A. E. Lugo, D. L. Graham, and W. D. Toone. 1994. Puerto Rican Parrots and potential limitations of the metapopulation approach to species conservation. Conserv. Biol. 8: 114-123.

Zeckhauser, R. J., and W. K. Viscusi. 1990. Risk within reason. Science 248: 559-564.

Jeffrey V. Wells: New York Cooperative Fish and Wildlife Research Unit, Fernow Hall, Cornell University, Ithaca, NY 14853 (current address: National Audubon Society, Cornell Laboratory of Ornithology, 159 Sapsucker Woods Road, Ithaca, NY 14850).

Use of Public Grazing Lands by Henslow's Sparrows, Grasshopper Sparrows, and Associated Grassland Birds in Central New York State

Charles R. Smith

Abstract

Habitat use by Henslow's Sparrows (*Ammodramus henslowii*), Grasshopper Sparrows (*A. savannarum*), and associated species of grassland birds was compared in pastures on Finger Lakes National Forest in central New York State. Henslow's Sparrows occupied pastures with significantly higher productivity and a slightly smaller percent of goldenrod (*Solidago* spp.) cover than pastures used by Grasshopper Sparrows. The average pasture size used by the two species did not differ significantly, but the smallest pasture containing Henslow's Sparrows (33 hectares [ha]) was significantly larger than the smallest pasture containing Grasshopper Sparrows (16 ha), suggesting greater area requirements for Henslow's Sparrows. Within territories, grass height was significantly greater for Henslow's Sparrows, but there was no significant difference in the percent cover of goldenrod. Among other grassland birds in the study area, Savannah Sparrows (*Passerculus sandwichensis*) accepted the widest range in pasture size and vegetation, and Bobolinks (*Dolichonyx oryzivorus*) were the most abundant birds.

The apparent need of Henslow's Sparrows for larger tracts of more productive land, and correspondingly taller vegetation, than Grasshopper Sparrows may partially explain why Henslow's Sparrows are much less widely distributed in New York than Grasshopper Sparrows; it may also account for the widespread decline of both species in the Northeast since the mid-1960s. Preservation of grasslands suitable for Henslow's Sparrows should also sustain Grasshopper Sparrows because such areas usually provide habitat for both species. Pasture management can maintain habitat suitable for both sparrows as well as for other grassland birds.

Introduction

In New York State and the Northeast, Grasshopper and Henslow's sparrows are uncommon grassland birds that breed in scattered locations. The New York State Breeding Bird Atlas reported the two species in 15 and 7 percent, respectively, of the blocks surveyed (Andrle and Carroll 1988). Grasshopper Sparrows have occurred in the Finger Lakes Region of New York since before 1909, having been described as "common" summer residents by Reed and Wright (1909). Henslow's Sparrows were not reported from the Finger Lakes Region by Reed and Wright (1909). Nonquantitative accounts have mentioned frequent decreases and increases in the abundance of these two species in New York (see regional reports for Region 3, Finger Lakes, in Kingbird 1955-1988). This is not surprising because populations of both species are reported to fluctuate widely, for reasons that are not understood (Hyde 1939, Wiens 1969, Zimmerman 1988).

U.S. Fish and Wildlife Service (USFWS) Breeding Bird Surveys (BBSs) have shown that Grasshopper and Henslow's sparrows are declining significantly in New York and throughout the Northeast (Robbins et al. 1986, Smith 1989, 1992, Smith and Smith 1992, Smith et al. 1993). Based on evaluations of long-term BBS data (Peterjohn et al. 1995), this regional decline appears to reflect a broader continental decline. Both species have been listed as either Blue List or Blue List special concern species since 1974 (Tate and Tate 1982, Tate 1986) and as New York State species of special concern since the early 1980s. Both species have continued to decline in New York, with Henslow's Sparrow showing a steeper decline than Grasshopper Sparrow. Over a broader geographic region, Vickery (1992) reported that Grasshopper Sparrow was listed as either endangered, threatened, or of special concern in all six New England states and New York and that Henslow's Sparrow was listed as extirpated in Connecticut and as endangered or of special concern in four of the other five New England states. In 1987 the USFWS listed Henslow's Sparrow as a migratory nongame bird of management concern (U.S. Fish Wildl. Serv. 1987).

The occurrence of both Grasshopper and Henslow's sparrows in pastures on Finger Lakes National Forest (NF; Fig. 1) in upstate New York provided an opportunity to study these birds and other grassland species on regularly grazed and managed grasslands. The primary goal of the study was to discern differences between the habitats of the two sparrow species and to make recommendations for their conservation. Based on preliminary observations, it was hypothesized that the territories of Henslow's Sparrows would have taller vegetation and a smaller percent cover of goldenrod than territories of Grasshopper

Grassland Birds in Central New York

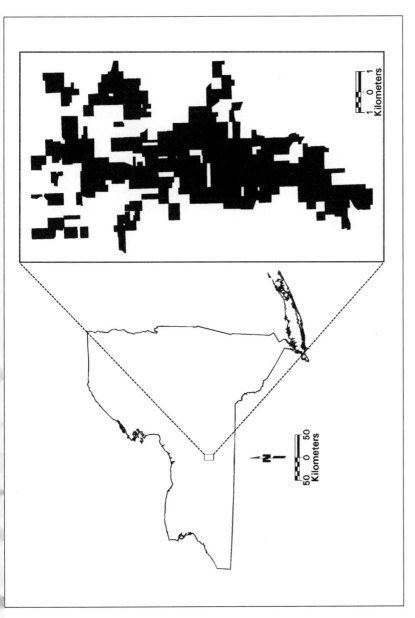

Figure 1. Map of New York State, showing location of Finger Lakes National Forest, on the boundary between Seneca and Schuyler counties. The 5,400-ha area is a patchwork of public lands (black shading), with some in-holdings of private lands (white areas surrounded by black). Most of the pastures at Finger Lakes National Forest are located in the northern half of its area. Map prepared by J. T. Weber, New York Gap Analysis Project.

Sparrows. It also was suspected that Henslow's Sparrows might be found in larger pastures and in pastures that had a higher productivity rating (as defined below; see "Study Area and Methods").

Habitats of Henslow's Sparrows and Grasshopper Sparrows

Henslow's Sparrows are found in a variety of habitats that contain tall, dense grass and herbaceous vegetation (Smith 1968). Hyde (1939) described several habitats where this species commonly occurs: upland weedy hayfields or pastures without shrubs, wet meadows, drier areas of salt marshes, grassy fields, and sedgy hillsides with recently planted pine (*Pinus*) seedlings. Graber (1968) found that the habitat of Henslow's Sparrows was usually quite dense from 30 to 61 centimeters (cm) off the ground and that the species "has adapted to living in unmowed hayfields," an observation confirmed by Bollinger (1995). In New York, Peterson (1983) found this sparrow in large, ungrazed fields, often on hilltops, with a variety of moisture regimes and no woody invasion. Wiens (1969) found that Henslow's Sparrow breeding territories had a low percent cover of forbs, dense vegetation, a high effective vegetation height, little bare ground or low vegetation, and no trees, posts, or fence lines. Henslow's Sparrows are not typically associated with grazed areas (Peterson 1983, Zimmerman 1988), although they often can survive well in pastures that are only lightly grazed (Skinner 1975).

Grasshopper Sparrows inhabit meadows, clover (*Trifolium* spp.) fields, and weedy fields, usually on drier ground than Henslow's Sparrows (Eaton 1914). On managed grasslands, Grasshopper Sparrows are most common in the presence of clump-forming vegetation, including alfalfa (*Medicago sativa*), red clover (*Trifolium pratense*), and lespedeza (*Lespedeza* spp.; Smith 1968). This preference for clump-forming vegetation has been confirmed by Whitmore (1981) and Janes (1983). Wiens (1969) found that Grasshopper Sparrow breeding territories frequently contained patches of bare ground and very short vegetation, posts, fence lines, and occasional trees. Johnston and Odum (1956) found the birds in fields with as much as 10 percent shrub cover, although more recent studies (Whitmore 1981, Janes 1983, CRS pers. obs.) have shown that shrub cover in breeding territories is usually closer to 1 percent. This apparently does not hold true for the Florida subspecies (*Ammodramus savannarum floridanus*), which may have territories with as much as 20 percent shrub cover (Delany et al. 1985).

Grassland Birds in Central New York

The effects on Grasshopper Sparrow populations of grazing differ from area to area. In Arizona, grazing reduced vegetation so much that Grasshopper Sparrows would not establish territories and breed (Bock and Webb 1984). In West Virginia (Whitmore 1981), North Dakota (Kantrud 1981, Renken and Dinsmore 1987), Florida (Delany et al. 1985), and Missouri (Skinner 1975), however, moderate grazing never decreased, and usually increased, the density of Grasshopper Sparrows. This difference may be explained by the preference of Grasshopper Sparrows in the West for fields with bunch grass interspersed with open ground (Janes 1983). In Arizona this habitat is found only in ungrazed areas because cattle quickly reduce the bunch grass to a density that is unacceptable to Grasshopper Sparrows. Farther east, light grazing creates patches of shorter vegetation usually not present in ungrazed grasslands, which encourages Grasshopper Sparrows. Heavy grazing can reduce the grass to an unacceptable height and density, causing declines in Grasshopper Sparrows but creating habitat suitable for other grassland species such as Vesper Sparrows (*Pooecetes gramineus*), which prefer shorter grass interspersed with patches of bare ground (Wiens 1969).

Study Area and Methods

Finger Lakes NF is the only land managed by the U.S. Forest Service (USFS) in New York State. Known as the Hector Land Use Area until 1985, the property was purchased between 1938 and 1941 as area farms failed during the Great Depression. The 5,400-ha area is now managed for multiple uses, including recreation, logging, and grazing. A variety of habitats are maintained by grazing, fire, and logging (Adkinson 1945, U.S. Forest Serv. 1986).

Thirty-three areas in Finger Lakes NF totaling about 735 ha and ranging in elevation from 330 to 560 meters (m) are designated as cattle pastures. These pastures vary in size from 8 to 124 ha, and each is fenced. Each year about 1,800 privately owned cattle graze the pastures from 15 May until 15 October. None of these fenced pastures comprise pure grassland; vegetation types within them can include as much as 36 percent shrubland and forest, as well as frequent tree and hedge rows. The pastures are stocked so that only 60 percent of the annual vegetation production is consumed by the cattle (U.S. Forest Serv. 1986). This results in large areas of tall grass that support populations of Eastern Meadowlarks (*Sturnella magna*), Bobolinks, and Savannah, Grasshopper, and Henslow's sparrows.

In the summer of 1989 I surveyed 33 pastures using two different procedures. From 26 to 29 June, I surveyed 11 pastures using line transects as suggested by Bollinger et al. (1988). From 10 to 21 July, I surveyed 22 pastures in a more general manner (see below). The transect method involved an afternoon visit to establish the transect, followed by an early morning visit to run the transect. The afternoon visit entailed establishing a line through the largest section of grassland and placing markers at 200-m intervals. The lengths of these transects ranged from less than 600 to more than 1,500 m, depending on the size of the pasture. I noted all bird species observed while establishing the transect. The early morning visit entailed walking each transect, stopping at each marker for five minutes, listing all species seen or heard, and counting all individuals seen and heard. The less intensive method used to survey 22 pastures involved simply walking in a large loop through each pasture and listing all species seen or heard, without counting individuals. I visited all 33 pastures predominantly during the morning hours, but the activity of the birds and the thoroughness of the technique allowed for detection of all grassland species, including Henslow's and Grasshopper sparrows, into the early afternoon.

From 26 July through 3 August, I visited the territories of 5 pairs of Henslow's Sparrows and 11 pairs of Grasshopper Sparrows to analyze vegetation. Observations on breeding territories were possible this late in the season because Grasshopper and Henslow's sparrows have a protracted breeding season that normally extends into the month of August (Smith 1968, Robins 1971a, b).

Data collection for the vegetation analysis involved the following procedure. Two observers located singing, territorial male birds. While one observer watched the bird, the other observer approached and caused the bird to fly. The observers then marked the original perch as well as where the flushed bird landed. This approach was similar to that employed by Wiens (1969) to establish territorial boundaries. With the original perch as the starting point, the observers laid a 50-m tape in the direction the flushed bird flew and measured the height of the vegetation at 1-m intervals along the tape. The actual stem measured for height was defined as the highest stem rooted within 10 cm of the original perch. Every 10 m the two observers placed a 1-m x 1-m square so that the square's diagonal was at right angles to the tape, and they visually estimated the percent cover of grass, goldenrod, sedge (*Carex*), and woody vegetation inside the square. They also counted goldenrod and other perennial stems, but goldenrod was the predominant forb in the study areas.

USFS staff provided a productivity index for each pasture. It was measured in head-months per hectare, or the number of cattle each pasture can support, times the number of months cattle are grazed on the pasture, divided by the number of hectares in the pasture. This productivity index is based on vegetation weight per unit area, estimated by clipping and weighing the new vegetation (omitting standing dead vegetation) in plots 89 cm square and randomly distributed throughout the pasture. The productivity index is updated yearly by reclipping and/or visual estimates (U.S. Forest Serv. 1977). This measure of standing vegetation, in weight per unit area, provides an estimate of vegetation density for each pasture, and this in turn is used by USFS staff to determine the number of cattle per hectare that each pasture can support.

For each pasture Finger Lakes NF staff also provided data on the total number of hectares of grassland, the percent forbs, and the date of the last mowing. Throughout the remainder of this paper, pasture size refers to the total amount of grassland in each pasture (obtained by subtracting the area of trees and shrubs from the total pasture area) and thus yields the amount of potentially usable habitat for Henslow's and Grasshopper sparrows.

I used Student's t-test to compare average grass heights on territories and average pasture sizes for Henslow's and Grasshopper sparrows. To ensure that the goldenrod data were normally distributed, I used the arcsin of the square root of the original data (Sokal and Rohlf 1981). I analyzed the productivity index for both species using a one-way analysis of variance (ANOVA).

Results

Grasshopper Sparrows were found in 24 of the 33 pastures and Henslow's Sparrows in 5. Three of the 33 pastures supported both species, and 7 of the 33 pastures supported neither (thus 21 of the 33 supported Grasshopper Sparrows only, and 2 of the 33 supported Henslow's Sparrows only). Of the five pastures with Henslow's Sparrows, one pasture had been set aside for hay and was not being grazed; four singing birds were present in a section of this nongrazed field that contained dense, tall grass with little goldenrod. This area was cut for hay in the third week of July, and the birds were not seen or heard again in any part of the field. Grasshopper Sparrows also nested in this field but used an area of shorter vegetation with more goldenrod. This area was not mowed, and the Grasshopper Sparrows completed their nesting undisturbed.

The average size of pastures containing Henslow's Sparrows (51.7 ha) was slightly greater than that of pastures containing Grasshopper Sparrows (43.4 ha), but the difference was not statistically significant (Table 1). The smallest pasture containing each species, or the minimum pasture size, was 16.2 ha for Grasshopper Sparrow and 33.2 ha for Henslow's Sparrow. The discovery of Grasshopper Sparrows on a

Table 1. *Summary of habitat differences in territories of Henslow's Sparrows (HESP) and Grasshopper Sparrows (GRSP) inhabiting pastures at Finger Lakes National Forest, NY, in 1989 (NS = not statistically significant, NA = not applicable).*

Habitat Attribute	HESP ($n = 5$)	GRSP ($n = 11$)	Significance
Mean grass height in territories (cm)	61.3	54.5	$p < 0.001$
Mean goldenrod (%) in territories	12.5	9.4	NS
Mean goldenrod/pasture (%)	25.0	38.5	NS
Mean grassland area (ha)	51.7	43.4	NS
Mean pasture productivity index (head-months/ha)	6.9	5.9	$p < 0.05$
Size (ha) of smallest grassland area containing each species	33.2	16.2	NA

16.2-ha pasture provides some insights into the minimum area required by this species, since this was the smallest area surveyed. The absence of Henslow's Sparrows in areas smaller than 33.2 ha may be more meaningful, because eight smaller pastures, including several with high productivity indexes, were surveyed and none contained Henslow's Sparrows. This tendency of Henslow's Sparrows to occupy larger pastures at Finger Lakes NF than Grasshoppper Sparrows continued through the 1995 field season.

The average productivity index for pastures containing Grasshopper Sparrows (5.9 head-months/ha) was significantly lower than for pastures containing Henslow's Sparrow (6.9 head-months/ha). The average height of the vegetation on the 5 Henslow's Sparrow territories (61.3 cm) was significantly greater than on the 11 Grasshopper Sparrow territories (54.5 cm). The height of the vegetation on the territories was significantly positively correlated ($p \leq 0.05$) with the productivity index of the pasture in which the territory was located.

Goldenrod cover was not significantly different on the territories of the two species. The very large variance of the results suggests that the sampling effort was not sufficient to determine if any real difference existed. Goldenrod cover per pasture (percent) was greater for pastures containing Grasshopper Sparrows, but this was not statistically significant. However, a p value of 0.09 suggests that this trend may be real and warrants further study.

Grasshopper and Henslow's sparrows did not seem to choose their breeding locations based on the time of last mowing, since both species bred freely on pastures mowed from one to six years earlier

(J. Fiske pers. comm.). For Henslow's Sparrows, two isolated territorial males and a colony of two or three territories were located in pastures mowed the previous year. The other two pastures supporting Henslow's Sparrows were mowed in 1984 and 1985. Grasshopper Sparrows were found in pastures that were last mowed in 1978 and in pastures mowed every year from 1982 to 1988.

Tables 2, 3, and 4 summarize attributes of the pastures on which Savannah, Grasshopper, and Henslow's sparrows, Bobolinks, and Eastern Meadowlarks established breeding territories. Savannah Sparrows showed the least specificity with respect to choice of pastures; they occurred on all the pastures at Finger Lakes NF, regardless of size or productivity, and comprised the second most abundant species occurring on a sample of 11 pastures for which relative

Table 2. Relationships among pasture size and productivity for five obligate grassland species occurring in pastures at Finger Lakes National Forest, NY, in June, July, and August 1989. Productivity is measured in head-months/hectare. Pastures used by these species ranged in size from 8 to 124 ha. Productivity indexes ranged from 4.0 to 8.1 head-months/ha.

Species	Smallest Pasture Used (ha)	Least Productive Pasture Used (ha)
Savannah Sparrow	11.7	4.0
Grasshopper Sparrow	16.2	4.0
Henslow's Sparrow	33.2	6.0
Bobolink	16.2	4.0
Eastern Meadowlark	23.5	4.8

Table 3. Multiple range analysis for pasture area (ha) by species. These five species constituted one homogeneous group (with respect to the average size of pasture occupied); no statistically significant differences were found among the five species.

Species	Sample Size	Average Size (ha)
Grasshopper Sparrow	24	49.1
Savannah Sparrow	33	53.6
Bobolink	30	56.5
Henslow's Sparrow	6	57.1
Eastern Meadowlark	19	61.7

Table 4. *Multiple range analysis for pasture productivity (head-months/ha) by species. These five species constituted two groups, with Henslow's Sparrow occupying significantly more productive pastures than the remaining four species. Sample size is the number of pastures sampled that were used by the species in 1989.*

Species	Sample Size	Average Productivity
Grasshopper Sparrow	24	5.9
Bobolink	30	6.0
Savannah Sparrow	33	6.1
Eastern Meadowlark	19	6.3
Henslow's Sparrow	6	6.8

abundances were estimated. Although not the most widespread species, Bobolinks were the most abundant of all the obligate grassland birds occurring at Finger Lakes NF. Grasshopper Sparrows were less common and occupied less productive pastures of a wide range of sizes. Eastern Meadowlarks ranked third among grassland birds in relative abundance and tended to occupy larger, more productive pastures, along with Henslow's Sparrows.

Discussion

The difference in mean grass heights between Grasshopper and Henslow's sparrows' territories supports the hypothesis that Henslow's Sparrows prefer taller vegetation than Grasshopper Sparrows. This result corroborates the findings of Skinner (1975) and Wiens (1969), who found Henslow's Sparrows inhabiting sites with taller vegetation than Grasshopper Sparrows. In Illinois, Herkert (1994a) also found Henslow's Sparrows associated with tall, dense vegetation.

The results with respect to the presence of goldenrod are less conclusive. Grasshopper Sparrows occurred in pastures with a higher percent goldenrod cover than Henslow's Sparrows, yet the two species did not show any difference between percent goldenrod cover within their territories. This apparent contradiction is reflected in the literature. Zimmerman (1988) and Whitmore (1981), working on Henslow's and Grasshopper sparrows, respectively, found that these species did not appear to choose territories based on the percent cover of forbs. But Wiens (1969) reported that Grasshopper Sparrow territories contained a larger percent of forbs than did Henslow's Sparrow territories. Unfortunately, the high variability of my within-territory data and the weak significance of the pasturewide data preclude resolution of these apparent contradictions.

In Kansas, Zimmerman (1988) found that Henslow's Sparrows did not breed in areas that had been burned the preceding spring or moderately grazed the preceding summer. Based on this, he concluded that any practice that reduced the standing dead vegetation in a field could eliminate Henslow's Sparrows. The occurrence of Henslow's Sparrows in central New York in grazed pastures that were mowed the previous year does not support Zimmerman's conclusion. The grazing intensities observed by Zimmerman may have been higher than those seen in this study, accounting for the absence of Henslow's Sparrows on his sites. The mowing discrepancy may be explained by one of two hypotheses. The first is that Zimmerman's work was done during June and mine during July and August; my later seasonal work gave the vegetation a chance to regrow and allowed the birds to move into these newly regrown areas after losing or raising their first brood. This could be the case since Robins (1971a) found that most Henslow's Sparrows in Michigan raise two or three broods, defend territories for as long as two months, and frequently change territory locations during the breeding season. The second hypothesis that may explain the discrepancy is that mowing during late July and August and removing the cattle in mid-October, as practiced at Finger Lakes NF, allows time for vegetation to regrow partially before winter, possibly providing enough residual cover in spring to attract Henslow's Sparrows. Also, the rate of vegetation growth probably is greater in New York, which gets more annual rainfall than Kansas.

Zimmerman (1988) proposed two explanations for the preference of Henslow's Sparrows for areas with large amounts of standing dead vegetation: first, that the standing dead vegetation discouraged new growth and provided more ground area for foraging; and second, that the standing dead vegetation protected the sparrows' nests from Brown-headed Cowbird (*Molothrus ater*) parasitism, predation, and microclimate extremes. Analysis of the productivity data from this study shows that Henslow's Sparrows prefer areas with a higher annual growth of new vegetation, suggesting that Zimmerman's first hypothesis is less applicable in the case of moderately grazed eastern pastures, since the Finger Lakes NF productivity index is based on annual new growth only (U.S. Forest Serv. 1977). In this study, where pastures were grazed annually and mowed periodically, there also was less opportunity for development of standing dead vegetation to which the birds might respond.

Kantrud (1981) found low Grasshopper Sparrow densities in hayfields mowed the previous year. Although I have no data on the relative densities of the birds in my study, the pattern of occurrence at Finger Lakes NF did not indicate that time since last mowing was important to choice of nesting location by Grasshopper Sparrows.

Two 33- to 35-ha pastures at Finger Lakes NF supported nesting Henslow's Sparrows, but this species was not found in smaller pastures. This result corresponds to the observations of Zimmerman (1988) in Kansas, who recommended that management to encourage Henslow's Sparrows be carried out on plots of at least 30 ha. Thirty-hectare pastures also fall within Samson's (1980) estimate of 10 to 100 ha as the minimum area required to support a viable breeding population of Henslow's Sparrows. In Illinois, Herkert (1994a) found Henslow's Sparrows only rarely in grasslands smaller than 100 ha. Peterson's (1983) study in Broome County, NY, showed that the occurrence of Henslow's Sparrows was related to distance from the horizon, a measure strongly correlated with grassland area. These apparent minimum area requirements may not hold for all regions where Henslow's Sparrows occur, however. Older accounts reported about 12 pairs living in 4 ha of dense grass in Pymatuning Swamp in northwestern Pennsylvania and 4 pairs in a field of only 3.6 ha (Graber 1968). The indications from recent work that size influences habitat choice by Henslow's Sparrow may be confounded by the fact that the species is declining. During periods of decline, a species is less likely to saturate the available habitats and may only occupy the highest quality sites (O'Connor 1981), giving a limited impression of the range of habitats it may occupy at higher population densities.

Conservation Implications

The current pasture management scheme at Finger Lakes NF, which has been in place since at least 1981, produces usable habitat for both Henslow's and Grasshopper sparrows. As long as this management plan continues, both species are likely to persist at Finger Lakes NF. Maintenance mowing should be done in early August to allow the birds to raise their first broods undisturbed, while still leaving enough time for regrowth to provide standing dead vegetation the following spring. This recommendation is consistent with that of Bollinger and Gavin (1992) for conservation of Bobolinks. Grazing is a cost-effective method in central New York for maintaining an early stage of succession, which undoubtedly benefits grassland birds. The stocking rate of only 0.12 to 0.24 head of cattle per hectare at Finger Lakes NF allows the vegetation to grow to the density required by Henslow's Sparrows. The presence of the largest colony of Henslow's Sparrows on the ungrazed area set aside for hay indicates that such areas provide good habitat for the birds. The only other ungrazed pasture at Finger Lakes NF, one with more than 80 ha of grassland and a low productivity index of 2.7, contained no Henslow's Sparrows, suggesting that both habitat area and habitat quality have to be considered in assessing the status of Henslow's Sparrow populations in New York and elsewhere.

The dairy industry in New York has been declining. The number of dairy cows in the state dropped from 914,000 in 1976 to 822,000 in 1988, which contributed to an overall decline in the number of cattle from 1,915,000 to 1,584,000 over the same time period (N.Y. State Agric. Stat. Serv. 1986, 1989). This decline has resulted in a decreased demand for pasturage at Finger Lakes NF. The fees charged for grazing rights help subsidize the cost of managing Finger Lakes NF (U.S. Forest Serv. 1986). If the number of cattle, and corresponding income from fees, are reduced, important management decisions will have to be made. Presently, grazing keeps the pastures in an early successional stage. With the loss of grazing, more frequent mowing will be necessary to maintain the grasslands. This will put additional stress on annual budgets and limited staff time and may not create the same habitat structure that grazing does. In the end, some pastures may need to be abandoned. With the 1976 mandate of the National Forest Management Act to preserve diversity on national forests (Bean 1983), managers at Finger Lakes NF face the challenge of maintaining a viable population of Henslow's Sparrows.

Conclusions

Productivity and minimum pasture size alone may not always accurately predict the presence or absence of breeding Henslow's Sparrows. This is not surprising, given that the species does not necessarily return to the same breeding sites each year (Wiens 1969, Robins 1971a, Peterson 1983), but these variables can help managers identify potential breeding sites. In his studies of midwestern grassland bird communities, Herkert (1994b) pointed out that both area and vegetation structure are important determinants of grassland bird species occurrence. By maintaining historical and potential breeding habitats, it should be possible to maintain Henslow's Sparrows at Finger Lakes NF.

The large number of pastures supporting Grasshopper Sparrows and the sizable number of individuals observed indicate that this species is doing well at Finger Lakes NF. The present management regime appears to be compatible with the habitat requirements of Grasshopper Sparrows, which do not readily frequent areas set aside for hay. This species' requirement for less productive land and smaller minimum areas, coupled with its relative abundance, seem to ensure its persistence. In Maine, at the northeastern limit of the species' breeding range, Vickery et al. (1994) reported that Grasshopper Sparrows reached an incidence of 50 percent occurrence on grasslands of 100 ha. In spite of this species' less demanding habitat requirements, the loss of grazing at Finger Lakes NF and elsewhere in the Northeast still could have negative effects on Grasshopper Sparrows, since some authors report that Grasshopper Sparrows are absent on ungrazed or idle pasture (Skinner 1975). The reasons for this discrepancy are uncertain, but grazing loss could reduce the suitability of the habitat, leading to a reduction in Grasshopper Sparrow densities. Another effect of reduced demand for pasturage at Finger Lakes NF could be a decision to allow some of the pastures to revert to trees and shrubs, also reducing the size of the Grasshopper Sparrow population at this site. Despite these potential decreases, preservation of suitable areas of high-quality grassland for Henslow's Sparrows should sustain Grasshopper Sparrows, because such areas usually can provide habitat for both species. By managing for the rarer Henslow's Sparrow, both species should remain a part of Finger Lakes NF's diverse avifauna.

Acknowledgments

The kind and enthusiastic assistance of the staff of the U.S. Forest Service, Finger Lakes National Forest, is gratefully acknowledged. W. P. Brown, W. Evans, E. D. O'Neill, and D. J. Smith provided valuable and dedicated assistance in the collection of field data. Financial support was provided by Hatch Projects NYC-171401 and NYC-147406, U.S. Department of Agriculture.

Literature Cited

Adkinson, L. B. 1945. Community use of pastures on hill lands: the Hector land use adjustment project in southern New York. U.S. Dept. Agric., Upper Darby, PA.

Andrle, R. F., and J. R. Carroll. 1988. The atlas of breeding birds in New York state. Cornell Univ. Press, Ithaca, NY.

Bean, M. J. 1983. The evolution of national wildlife law. Praeger, New York.

Bock, C. E., and B. Webb. 1984. Birds as grazing indicator species in southeastern Arizona. J. Wildl. Manage. 48: 1045–1049.

Bollinger, E. K. 1995. Successional changes and habitat selection in hayfield bird communities. Auk 112: 720–730.

Bollinger, E. K., and T. A. Gavin. 1992. Eastern Bobolink populations: ecology and conservation in an agricultural landscape. Pp. 497–506 *in* Ecology and conservation of neotropical migrant landbirds (J. M. Hagan III and D. W. Johnston, eds.). Smithsonian Inst. Press, Washington, D.C.

Bollinger, E. K., T. A. Gavin, and D. C. McIntyre. 1988. Comparison of transects and circular-plots for estimating Bobolink densities. J. Wildl. Manage. 52: 777-786.

Delany, M. F., H. M. Stevenson, and R. McCracken. 1985. Distribution, abundance and habitat of the Florida Grasshopper Sparrow. J. Wildl. Manage. 49: 626-631.

Eaton, E. H. 1914. Birds of New York. Pt. 2. Univ. of the State of New York, Albany.

Graber, J. W. 1968. *Passerherbulus henslowii henslowii* Western Henslow's Sparrow. Pp. 779-788 *in* Life histories of North American cardinals, grosbeaks, buntings, towhees, finches, sparrows and allies (A. C. Bent, ed.). U.S. Natl. Mus. Bull. 237.

Herkert, J. R. 1994a. Status and habitat selection of the Henslow's Sparrow in Illinois. Wilson Bull. 106: 35-45.

Herkert, J. R. 1994b. The effects of habitat fragmentation on midwestern grassland bird communities. Ecol. Appl. 4: 461-471.

Hyde, A. S. 1939. The life history of Henslow's Sparrow (*Passerherbulus henslowii*). Mus. Zool. Univ. Mich. Misc. Publ. no. 41.

Janes, J. W. 1983. Status, distribution and habitat selection of Grasshopper Sparrows in Morrow County, Oregon. Murrelet 64: 51-54.

Johnston, D. W., and E. P. Odum. 1956. Breeding bird populations in relation to plant succession on the Piedmont of Georgia. Ecology 38: 171-174.

Kantrud, H. A. 1981. Grazing intensity effects on the breeding avifauna of North Dakota native grasslands. Can. Field-Nat. 95: 404-417.

New York State Agricultural Statistics Service. 1986. New York Agricultural Statistics 1985. N. Y. State Dept. Agric. and Markets, Div. Statistics, Albany.

New York State Agricultural Statistics Service. 1989. New York Agricultural Statistics 1988-1989. N. Y. State Dept. Agric. and Markets, Div. Statistics, Albany.

O'Connor, R. J. 1981. Habitat correlates of bird distribution in British census plots. Stud. Avian Biol. 6: 533-537.

Peterjohn, B. J., J. R. Sauer, and S. Orsillo. 1995. Breeding bird survey: population trends 1966-92. Pp. 17-21 *in* Our living resources: a report to the nation on the distribution, abundance, and health of U.S. plants, animals, and ecosystems (E. T. LaRoe, G. S. Farris, C. E. Puckett, P. D. Doran, and M. J. Mac, eds.). U.S. Dept. Interior, Natl. Biol. Serv., Washington, D.C.

Peterson, A. 1983. Observations on habitat selection by Henslow's Sparrow in Broome County, New York. Kingbird 33: 155-164.

Reed, H. D., and A. H. Wright. 1909. The vertebrates of the Cayuga Lake basin, N.Y. Proc. Am. Phil. Soc. 48: 370-459.

Renken, R. B., and J. J. Dinsmore. 1987. Nongame bird communities on managed grasslands in North Dakota. Can. Field-Nat. 101: 551-557.

Robbins, C. S., D. Bystrak, and P. H. Geissler. 1986. The breeding bird survey: its first fifteen years, 1965-1979. U.S. Fish Wildl. Serv. Res. Publ. 157.

Robins, J. D. 1971a. A study of Henslow's Sparrow in Michigan. Wilson Bull. 88: 39-48.

Robins, J. D. 1971b. Differential niche utilization in a grassland sparrow. Ecology 52: 1065-1070.

Samson, F. B. 1980. Island biogeography and the conservation of nongame birds. Trans. N. Am. Wildl. Nat. Res. Conf. 45: 245-251.

Skinner, R. M. 1975. Grassland use patterns and prairie bird populations in Missouri. Pp. 171-180 *in* Prairie: a multiple view (M. K. Wali, ed.). Univ. of N. Dakota Press, Grand Forks.

Smith, C. R. 1989. An analysis of New York state breeding bird surveys 1966-1985. Final project report, contract no. C001667, N. Y. State Dept. Env. Conserv., Ithaca.

Smith, C. R. 1992. Henslow's Sparrow. Pp. 315-330 *in* Migratory nongame birds of management concern in the northeast (K. J. Schneider and D. M. Pence, eds.). U.S. Fish Wildl. Serv., Newton Corner, MA.

Smith, C. R., D. M. Pence, and R. J. O'Connor. 1993. Status of neotropical migratory birds in the northeast: a preliminary assessment. Pp. 172-188 *in* Status and management of neotropical migratory birds (D. M. Finch and P. W. Stangel, eds.). Gen. Tech. Rep. RM-229, U.S. Forest Serv., Rocky Mtn. Forest Range Exper. Stn., Fort Collins, CO.

Smith, D. J., and C. R. Smith. 1992. Henslow's Sparrow and Grasshopper Sparrow: a comparison of habitat use in Finger Lakes National Forest, New York. Bird Observer 20: 187-194.

Smith, W. P. 1968. Eastern Henslow's Sparrow. Pp. 776-778 *in* Life histories of North American cardinals, grosbeaks, buntings, towhees, finches, sparrows and allies (A. C. Bent, ed.). U.S. Natl. Mus. Bull. 237.

Sokal, R. R., and F. J. Rohlf. 1981. Biometry. 2d ed. W. H. Freeman, New York.

Tate, J. 1986. The blue list for 1986. Am. Birds. 40: 227-236.

Tate, J., and J. Tate. 1982. The blue list for 1982. Am. Birds. 36: 126-135.

U.S. Fish and Wildlife Service. 1987. Migratory nongame birds of management concern in the United States: the 1987 list. Office Migratory Bird Manage., U.S. Fish Wildl. Serv., Washington, D.C.

U.S. Forest Service. 1977. Herbage/forage production estimates. U.S. Forest Serv. range analysis and management handbook 2209.14. U.S. Forest Serv., Milwaukee, WI.

U.S. Forest Service. 1986. Land and resource management plan: Finger Lakes National Forest. U.S. Forest Serv., Rutland, VT.

Vickery, P. D. 1992. A regional analysis of endangered, threatened, and special-concern birds in the northeastern United States. Trans. N. E. Sec. Wildl. Soc. 48: 1-10.

Vickery, P. D., M. L. Hunter, Jr., and S. M. Melvin. 1994. Effects of habitat area on the distribution of grassland birds in Maine. Conserv. Biol. 8: 1087-1097.

Whitmore, R. C. 1981. Structural characteristics of Grasshopper Sparrow habitat. J. Wildl. Manage. 45: 811-814.

Wiens, J. A. 1969. An approach to the study of ecological relationships among grassland birds. Ornithol. Monogr. 8.

Zimmerman, J. L. 1988. Breeding season habitat selection by the Henslow's Sparrow in Kansas. Wilson Bull. 100: 17-24.

Charles R. Smith: Department of Natural Resources and New York Cooperative Fish and Wildlife Research Unit, Fernow Hall, Cornell University, Ithaca, NY 14853-3001.

Distribution and Population Status of Grassland Birds in Massachusetts

Andrea L. Jones and Peter D. Vickery

Abstract

In 1993 the Center for Biological Conservation at the Massachusetts Audubon Society initiated a three-year study of the distributions and populations of nesting grassland birds in Massachusetts. The survey focused on three species: Upland Sandpiper (*Bartramia longicauda*), Vesper Sparrow (*Pooecetes gramineus*), and Grasshopper Sparrow (*Ammodramus savannarum*). It also included Savannah Sparrow (*Passerculus sandwichensis*), Bobolink (*Dolichonyx oryzivorus*), and Eastern Meadowlark (*Sturnella magna*). During three nesting seasons (1993-1995) we recorded approximately 80 pairs of Upland Sandpipers, 130 territorial male Vesper Sparrows, and 350 territorial male Grasshopper Sparrows in the state. Because we documented such low numbers of Vesper and Grasshopper sparrows, both species have been state listed as threatened by the Massachusetts Division of Fisheries and Wildlife, based on recommendations made by the Center for Biological Conservation. Upland Sandpiper had already been state listed as endangered in 1985, and it remains so. The Center for Biological Conservation is currently working on management recommendations to reverse the declines of these grassland species.

A. L. Jones and P. D. Vickery

Introduction

Because of well-documented declines in grassland bird populations throughout northeastern North America since 1965 (Vickery 1992, Askins 1993), in 1993 the Center for Biological Conservation initiated a study of the distributions and populations of nesting grassland birds in Massachusetts. The survey focused on three regionally rare species, Upland Sandpiper and Vesper and Grasshopper sparrows, but also included Savannah Sparrow, Bobolink, and Eastern Meadowlark. Researchers, naturalists, and birders throughout the state helped us identify promising grassland tracts, and during the 1993–1995 field seasons we counted populations at grassland sites from the Berkshire Mountains to Cape Cod and on the islands south of Cape Cod. The data we gathered have provided important insights into the continuing story of the rise and fall of grassland bird populations in New England.

Loss of habitat appears to be the primary reason for the decline of grassland birds in the Northeast (Askins 1993, Vickery et al. 1994). Grassland birds are now reduced to breeding at airports, remnant hayfields and pastures, and meadows in conservation areas. The majority of grassland sites in Massachusetts are currently concentrated in the Connecticut River valley and on offshore islands, especially Martha's Vineyard, Nantucket, and the Elizabeth Islands. However, continuing suburban and industrial development, with consequent habitat loss and fragmentation, places additional constraints on these habitats to support breeding populations of grassland birds.

Nearly all grassland birds have specific habitat requirements for vegetation type and size of grassland area. Most grassland birds require large patches of contiguous habitat for breeding; small, fragmented grasslands are too small to support grassland birds. Upland Sandpipers, for example, are rare on sites smaller than 50 hectares (ha) and become more common on sites of about 200 ha. Vesper Sparrows prefer areas of approximately 20 ha, and Grasshopper Sparrows prefer areas of at least 80 ha (Vickery et al. 1994). Eastern Meadowlarks, Bobolinks, and Savannah Sparrows are less dependent on area and therefore still breed in many smaller fields throughout the state. Grasslands need to comprise at least 200 ha to maintain a diverse population of grassland bird species (Vickery et al. 1994). Although some large-scale monoculture farms still exist in New England, especially in Vermont, these are usually unsuitable for grassland birds such as Upland Sandpiper, which require a mosaic of grassland habitats and grass lengths.

Native grasslands occurred sporadically in New England in pre-Colonial times. These grasslands were maintained by fire, either natural, or more frequently, set by Native Americans (Patterson and Sassaman 1988). These open areas provided habitat for grassland birds, including the endemic Heath Hen (*Tympanuchus cupido cupido*), a subspecies of the Greater Prairie-Chicken that became extinct in the 1930s. Because fires are now suppressed, many native grasslands and heathlands have changed into thick shrublands or forests that are unsuitable for grassland birds.

There is little doubt that grassland bird populations in Massachusetts reached a zenith in the late eighteenth and early nineteenth centuries as eastern forests were cut and converted to large farms. Since the late 1800s, however, more than 60 percent of New England's farms have disappeared, and forests have reclaimed many of the fields that once supported such species as Vesper Sparrows and Upland Sandpipers (Litvaitis 1993, Vickery et al. 1994). Some grasslands are no longer maintained, and as trees and shrubs have encroached on them, the area and quality of these grassland habitats have been further diminished. In addition, modern agricultural practices, such as early and more frequent summer mowing of hayfields, can destroy nests before young have fledged. Because of frequent mowing at most airports, birds often are unable to establish successful nests or are exposed to predators.

Interestingly, since the late 1980s large (approximately 20 ha or larger) landfills have become breeding habitats for some grassland birds. As landfills reach full capacity, they are capped with plastic shields that cover the fill material. These shields are topped with 45 centimeters (cm) of soil and are seeded with grass. These dry grasslands must be mowed every year to ensure that the roots of shrubs or trees will not penetrate the plastic cap. Ironically, since the late 1980s several of these artificial grasslands (e.g., in Worcester and Clinton, MA) have provided breeding habitat for Grasshopper and Savannah sparrows, Bobolinks, and Eastern Meadowlarks.

Status of Regionally Threatened Grassland Birds in Massachusetts

Upland Sandpiper

In the mid-nineteenth century, Upland Sandpipers were common; "breeds, and towards autumn is often very common" wrote J. A. Allen (*in* Bagg and Eliot 1937) in 1864. But by the turn of the century, this species was declining, and it was soon noted that "this once well-represented tattler has come close to extinction in our region during the past 30 years" (Bagg and Eliot 1937).

The Upland Sandpiper has declined rapidly to the point of becoming state listed in all seven states in the New England–New York region

(Vickery 1992). It is listed as endangered in two states (including Massachusetts), threatened in three, and of special concern in two (Vickery 1992). Breeding areas are currently restricted to a few airports, military bases, and one remaining large expanse of farmland in southeastern Massachusetts. The species is now considered "local and very uncommon...greatly decreased since the 1800s" (Veit and Petersen 1993).

Vesper Sparrow

Vesper Sparrows were common in Massachusetts during the late nineteenth and early twentieth centuries, particularly along the Connecticut River valley and on Cape Cod, inhabiting dry farmlands and sandy areas. The species was noted as a "common summer resident" (Bagg and Eliot 1937) and in 1864 was "abundant breeding in open sandy fields and dry pastures" (J. A. Allen *in* Bagg and Eliot 1937). A decline became evident by the mid-twentieth century, however, at which time the species was noted as "formerly an abundant summer resident in open farming country throughout the state, now rapidly decreasing and becoming rare and local with the decline of agriculture" (Griscom and Snyder 1955).

Vesper Sparrow populations have declined sharply throughout New England. Within the New England–New York area, the species is listed as state endangered in two states, threatened in one (Massachusetts), and of special concern in three (Vickery 1992). In Massachusetts it is now considered an "uncommon and local breeder" (Veit and Petersen 1993).

Grasshopper Sparrow

During the late nineteenth and early twentieth centuries, Grasshopper Sparrows were abundant, particularly in the Connecticut River valley, on Cape Cod, and on the islands; "in our central valley, the Grasshopper Sparrow, though little known, has long been, locally, abundant....[In] the wide fields along the Amherst/Hadley line it is now far from rare" (Bagg and Eliot 1937). Within a few decades, however, a decline was documented: "formerly an abundant summer resident (Cape Cod, Nantucket, Martha's Vineyard) occurring more locally north to Essex County and in the two inland river valleys. It has greatly decreased and is now becoming rare and local" (Griscom and Snyder 1955).

The Grasshopper Sparrow is currently state listed by all seven states in the New England–New York area. It is listed as endangered in two states, threatened in two (including Massachusetts), and of special concern in three (Vickery 1992). The only large populations of more than 100 pairs include Westover Air Reserve Base (ARB) in the western part of the state and Nashawena Island in the Elizabeth Islands. The species is now considered a "rare to uncommon breeder" (Veit and Petersen 1993).

Survey Methods

We concentrated our survey efforts on the three species we considered to be in the greatest jeopardy: Upland Sandpiper and Grasshopper and Vesper sparrows. Ornithologists and naturalists throughout the state helped us identify historical or current grassland bird nesting areas, and we conducted a statewide survey at 150 sites during the 1993-1995 field seasons (Fig. 1). These sites included small hayfields, farm pastures, private and public airports, and large military air bases. Locations ranged from the Berkshire Mountains to Provincetown and included the Elizabeth Islands, Martha's Vineyard, and Nantucket.

We conducted surveys between 15 May and 15 June using a point census technique with a 100-meter (m) radius. For five minutes we recorded all birds we saw or heard. We then played a tape recording of Vesper Sparrow, Grasshopper Sparrow, Upland Sandpiper, Savannah Sparrow, Eastern Meadowlark, and Bobolink songs; each species' song

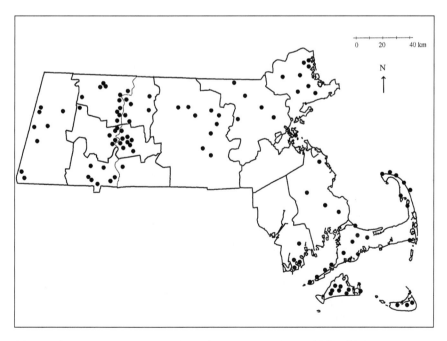

Figure 1. Massachusetts grassland bird survey sites, 1993-1995.

was approximately 30 seconds long with a 30-second interval between species. We counted birds that responded to the tape, as well as birds we saw or heard while walking between census points. Depending on the size of the grassland, we censused three to five points at each site. On larger and more productive sites, we conducted a second survey, usually two weeks (or sooner) after the first survey. At these sites, a complete walk-through of the grasslands, while we periodically played the tape recording, gave a more precise census of grassland bird numbers. Additional surveys were conducted by Massachusetts Division of Fisheries and Wildlife staff at Fort Devens (Ayer), Westover ARB (Chicopee), Camp Edwards Military Reservation (Sandwich), and Otis Air Force Base Airfield (Sandwich).

Results

We located a total of 53 sites that supported one or more species of regionally threatened or endangered nesting grassland birds (see Fig. 2, Table 1, and Appendix).

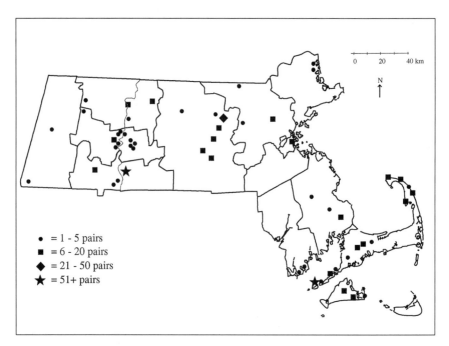

Figure 2. *Sites supporting one or more species of regionally threatened or endangered nesting grassland birds (Upland Sandpiper, Vesper Sparrow, and Grasshopper Sparrow) in Massachusetts in 1993–1995; 53 sites are denoted.*

Status of Grassland Birds in Massachusetts

Table 1. Numbers of grassland birds counted during 1993-1995 field seasons at 53 sites in Massachusetts. Number for Upland Sandpiper indicates approximate number of adult pairs. Numbers for Vesper Sparrow, Grasshopper Sparrow, Eastern Meadowlark, Savannah Sparrow, and Bobolink indicate singing males.

Species	Numbers	Number of Sites
Upland Sandpiper	80	10
Vesper Sparrow	130	27
Grasshopper Sparrow	350	32
Eastern Meadowlark*	260	40
Bobolink*	700	55
Savannah Sparrow*	535	67

*Numbers for Eastern Meadowlark, Bobolink, and Savannah Sparrow do not represent complete statewide population numbers.

Upland Sandpiper

We counted approximately 80 pairs of Upland Sandpipers at 10 sites. Only four sites had more than four pairs: Logan Airport (Boston), Hanscom Field (Bedford), Camp Edwards Military Reservation (Sandwich), and Westover ARB (Chicopee). The other six sites supported only one to three breeding pairs. At Westover ARB, the only site actively managed for this species, the population increased from approximately 20 breeding pairs in 1986 to approximately 50 in 1994 (Melvin 1994).

A major reason for the decline in Upland Sandpipers in Massachusetts and elsewhere in the Northeast is the amount of breeding habitat this species requires: at least 100 ha of optimal habitat, with surrounding open country, is preferred (Vickery et al. 1994). As farms have become smaller and more fragmented since the late 1800s, expansive open habitats no longer exist in most locations and few large grasslands remain.

Vesper Sparrow

We counted only 130 singing male Vesper Sparrows at 27 sites (see Appendix). The following sites had more than four singing males: Plymouth Airport (Plymouth), Fort Devens parachute landing (Ayer), Orange Airport (Orange), Hadley Honey Pot Farm (Hadley), Bashin Road (Hatfield), Provincetown Sand Dunes (Truro), Provincetown Airport (Provincetown), Marconi Barrens, Cape Cod National Seashore (Wellfleet), Griffin's Island (Wellfleet), and Northampton Meadows

(Northampton). Approximately 21 singing males scattered throughout the potato fields on the Bashin Road area in Hatfield, bordering the Connecticut River, formed the largest population in the state. The Provincetown Sand Dunes and Marconi Barrens, both on Cape Cod, supported other sizable breeding populations, each with 10 singing males. The remaining birds were scattered across the state in small populations.

Vesper Sparrows are now confined primarily to a few large agricultural areas in the Connecticut River valley, the protected coastal moors of Cape Cod, and small airports. Unlike Grasshopper Sparrows, Vesper Sparrows do not occur in any large concentrations in Massachusetts; they are more widely distributed than Grasshopper Sparrows, but only a few pairs have been found at each site. Most of these sites are unmanaged, and it is uncertain whether they will continue to provide suitable breeding habitat. Although breeding sites on Cape Cod are protected by the Cape Cod National Seashore, the number of Vesper Sparrows at Marconi Barrens has decreased since the mid-1980s as shrubs have encroached on open habitat. Without active management, it is unlikely that this species will persist at this site.

Because Vesper Sparrow populations have declined sharply since the late 1800s, in 1994 the Center for Biological Conservation recommended that the species be state listed as endangered or threatened. It was state listed as threatened in 1995.

Grasshopper Sparrow

We counted approximately 350 singing male Grasshopper Sparrows at 32 sites (see Appendix). The following seven sites had eight or more singing males: Worcester landfill (Worcester), Fort Devens parachute landing (Ayer), Westover ARB (Chicopee), Turner's Falls Airport (Turners Falls), Katama Airfield (Martha's Vineyard), Dukes County Airport (Martha's Vineyard), and Nashawena Island. More than 75 percent of the state's Grasshopper Sparrow population occurred at only two sites: Nashawena Island and Westover ARB. The largest population in Massachusetts, on Nashawena Island, declined by 58 percent (to 125 singing males) from 1988 to 1994 (P. Vickery unpubl. data), whereas the population at Westover ARB increased from 55 to 168 singing males during the same period (Melvin 1994).

Because there has been a major, long-term decline across most of the state, and because such a high proportion of the state's population is limited to just two sites, one of which is declining precipitously, in 1994 the Center for Biological Conservation recommended that the Grasshopper Sparrow be state listed as endangered or threatened. It was state listed as threatened in 1995.

Savannah Sparrow, Bobolink, and Eastern Meadowlark

We found each of these species scattered throughout Massachusetts, particularly in the Connecticut River valley, where the majority of the state's farmland still exists. Each species has adapted to a variety of agricultural situations and, unlike the Upland Sandpiper and Grasshopper Sparrow, none has a large area requirement for breeding (Vickery et al. 1994). However, because of a variety of factors, the breeding populations of these species have declined throughout Massachusetts and northeastern North America (Askins 1993, Veit and Petersen 1993). Bobolinks, for instance, which nest primarily in hayfields, are threatened by early mowing practices. In Lincoln, MA, an 80 percent mortality of Bobolink young was found in active hayfields mowed throughout the breeding season (Ells 1995). The Center for Biological Conservation will continue surveys in the Connecticut River valley to document further breeding sites for these species. Currently, none of these species is state listed as threatened or endangered.

Discussion

Our inventory of 150 sites revealed that some of the best areas for nesting grassland birds in Massachusetts are military airfields. The large grasslands surrounding the runways provide high-quality breeding habitat for many species. Westover ARB in Chicopee, with about

620 ha of grassland habitat, is presently the state's most important site for nesting grassland birds (Melvin 1994). The Massachusetts Natural Heritage and Endangered Species Program has worked closely with environmental engineers on this base to develop management practices that are compatible with grassland bird nesting ecology. Except for areas immediately adjacent to runways, mowing is deferred until August, after the nesting season. Should Westover ARB or the other military sites in Massachusetts close, it is likely that they would no longer be maintained as grasslands and that breeding Upland Sandpipers and other grassland birds would disappear rapidly.

Management

The best opportunity to protect and enhance nesting grassland bird populations is through better habitat management of existing grassland sites. If mowing can be delayed on federal, state, and municipal lands until the end of the breeding season (approximately 31 July), reproductive success improves dramatically. Active grassland-habitat management at Westover ARB has demonstrated that grassland birds respond positively to deferred mowing. Conversely, on the Elizabeth Islands, where grassland habitat has not been carefully managed, the grasslands have reverted to shrubland on several islands, and Grasshopper Sparrow populations have declined by more than 40 percent since 1985 (P. Vickery unpubl. data).

Prescribed burning has been demonstrated to improve habitat for most grassland birds (Vickery 1993) and has been an important management technique used by the Massachusetts Audubon Society on Nantucket Island and throughout Massachusetts. To date, however, only a few grassland sites have been burned to improve the bird habitat.

The Center for Biological Conservation produced several grassland-management manuals that are being distributed to landowners throughout New England. This literature provides management options for all grassland flora and fauna. By using these management tools and by educating private, municipal, state, and federal land-managers about these practices, we hope to halt and reverse further declines of these rare birds.

Acknowledgments

We are grateful to the many birders who provided historical information and suggestions on possible grassland bird breeding locations and to the many people who helped conduct surveys. We are especially thankful for the field assistance provided by M. Eckman,

S. Perkins, W. Petersen, and J. Rivers. Staff at the Arcadia Wildlife Sanctuary, Wellfleet Bay Wildlife Sanctuary, and Hitchcock Center for the Environment provided logistical support, access to lands, and volunteers. We thank the many landowners across Massachusetts for permitting access to their lands. We thank the Forbes family, The Trustees of Reservations, and The Nature Conservancy for permitting access to sites on the Elizabeth Islands and Martha's Vineyard; the Cape Cod National Seashore for providing transportation and access to the Provincelands; and N. Smith of Blue Hills Trailside Museum for helping us gain access to Logan Airport. S. M. Melvin, of the Massachusetts Division of Fisheries and Wildlife's Endangered Species and Natural Heritage Program, provided survey data at military reservations and facilitated access to private and municipal airports. Our research was supported in part by grants from the Switzer Environmental Fellowship Program and the George H. and Jane A. Mifflin Memorial Fund.

Literature Cited

Askins, R. A. 1993. Population trends in grassland, shrubland, and forest birds in eastern North America. Current Ornithol. 11: 1-34.

Bagg, A. C., and S. A. Eliot, Jr. 1937. Birds of the Connecticut valley in Massachusetts. Hampshire Bookshop, Northampton, MA.

Ells, S. 1995. Bobolink protection and mortality on suburban conservation lands. Bird Observer 23: 98-112.

Griscom, L., and D. D. Snyder. 1955. The birds of Massachusetts. Peabody Mus., Salem, MA.

Litvaitis, J. A. 1993. Response of early successional vertebrates to historic changes in land use. Conserv. Biol. 7: 866-873.

Melvin, S. M. 1993. 1993 census results and management recommendations for grassland birds at Westover Air Reserve Base, Massachusetts. Mass. Div. Fish. Wildl., Boston.

Melvin. S. M. 1994. Military bases provide habitat for rare grassland birds. Mass. Div. Fish. Wildl. Nat. Heritage News 4: 3.

Patterson, W. A., III, and K. E. Sassaman. 1988. Indian fires in the prehistory of New England. Pp. 107-135 *in* Holocene human ecology in northeastern North America (G. P. Nicholas, ed.). Plenum Press, New York.

Veit, R. R., and W. R. Petersen. 1993. Birds of Massachusetts. Massachusetts Audubon Soc., Lincoln.

Vickery, P. D. 1992. A regional analysis of endangered, threatened, and special-concern birds in the northeastern United States. Trans. N. E. Sect. Wildl. Soc. 48: 1-10.

Vickery, P. D. 1993. Habitat selection of grassland birds in Maine. Ph.D. diss., Univ. of Maine, Orono.

Vickery, P. D., M. L. Hunter, Jr., and S. M. Melvin. 1994. Effects of habitat area on the distribution of grassland birds in Maine. Conserv. Biol. 8: 1087-1097.

A. L. Jones and P. D. Vickery

Andrea L. Jones: Center for Biological Conservation, Massachusetts Audubon Society, Lincoln, MA 01773.

Peter D. Vickery: Center for Biological Conservation, Massachusetts Audubon Society, Lincoln, MA 01773, and Department of Forestry and Wildlife Conservation, University of Massachusetts, Amherst, MA 01003.

Appendix. List of survey sites in Massachusetts for Upland Sandpipers, Vesper Sparrows, and Grasshopper Sparrows, 1993–1995.

Upland Sandpiper: 1993–1995

Site (Total Number Sites = 10)	Number pairs		
	1993	1994	1995
Logan Airport, Boston	5		9
Hanscom Field, Bedford	4		5
Plum Island Airport, Newburyport	1	1	1
Cumberland Farm, Middleboro	4	4	4
Fort Devens parachute landing, Ayer	1	1	2
Fort Devens capped landfill, Ayer	1		
Westover Air Reserve Base, Chicopee*	25	50	40
Barnes Airport, Westfield	1		
Camp Edwards Military Reservation, Sandwich*	8		
Otis Air Force Base Airfield, Sandwich*	3		
Total	53	56	61

Total Population Estimate, 1993–1995: Approximately 80 pairs.

Vesper Sparrow: 1993–1995

Site (Total Number Sites = 27)	Number singing males		
	1993	1994	1995
Miles Standish State Forest, Plymouth	2		
Plymouth Airport, Plymouth	5	4	
Gardner Airport, Gardner	2		
Fort Devens parachute landing, Ayer	7		3
Puffer's gravel pit, Leverett		2	
Agawam Industrial Park, Agawam	4		
Barnes Airport, Westfield	3		
Orange Airport, Orange	8		
Turners Falls Airport, Turners Falls	3		2
Hadley Honey Pot Farm, Hadley		2	8
Bull Hill Road fields, Sunderland	2		2
Bashin Road potato fields, Hatfield	1		21
Pilgrim Heights National Seashore, Truro	1		
Provinceland Sand Dunes, Truro		10	
Provincetown Airport, Provincetown	7		
Marconi Barrens, Cape Cod National Seashore, Wellfleet	10	9	11
East Road fields, Pepperell	1		

Status of Grassland Birds in Massachusetts

Site	1993	1994	1995
High Street fields, Attleboro			3
Northampton Meadows, Northampton	3		7
Bowles fields, Agawam		1	
South Meadows, Deerfield		1	
Potato field, Plainfield	2	2	
Potato field, Hawley	4	4	
Potato field, Worthington	3	3	
Griffin's Island, Wellfleet		9	3
Otis Air Force Base Airfield, Sandwich*	2		
Westover Air Reserve Base, Chicopee*	3	2	2
Total	73	49	62

Total Population Estimate, 1993–1995: Approximately 130 singing males.

Grasshopper Sparrow: 1993–1995

Site (Total Number Sites = 32)	Number singing males		
	1993	1994	1995
Logan Airport, Boston	3		
Hanscom Field, Bedford	5		1
Delaney Wildlife Management Area, Stow	2	1	
Barney's Joy Farm, S. Dartmouth	2		2
Gardner Airport, Gardner	2		
Worcester landfill		8	5
Worcester Municipal Airport, Worcester*	6	2	2
Fort Devens parachute landing, Ayer	17		22
Fort Devens capped landfill, Ayer	2		
Clinton landfill, Clinton	4	7	
Westover Air Reserve Base, Chicopee*	99	168	150+
Agawam Industrial Park, Agawam	1		
Barnes Airport, Westfield	5		
Orange Airport, Orange	1		
Turners Falls Airport, Turners Falls	8		12
Bull Hill Road fields, Sunderland			3
Amherst landfill, Amherst		2	
Crane Wildlife Management Area, Falmouth	2	2	6
Katama Airfield, Martha's Vineyard	5	9	8
Dukes County Airport, Martha's Vineyard		8	5
Naushon Island	7		
Nonamesset Island	4		
Pasque Island	2		
Nashawena Island	125	125	115
Plymouth Airport, Plymouth	1		
Northampton State Hospital, Northampton	1		
Burts Pit Road fields, Northampton	6		
Kaiser Permanente fields, Amherst		2	
Brookfield Trail Conservation Area, Amherst		1	
Bowles fields, Agawam		1	
Sheffield fields, Sheffield		1	
E. Elm Street fields, W. Bridgewater			4
Total	310	337	335+

Total Population Estimate, 1993–1995: Approximately 350 singing males.

*Survey by Massachusetts Division of Fisheries and Wildlife

Grassland Birds in Vermont: Population Status, Conservation Problems, and Research Needs

Charles Darmstadt, Christopher Rimmer, Judith Peterson, and Christopher Fichtel

Introduction

Grassland bird species have received little scientific study in Vermont. Most research and monitoring efforts have been limited to the Champlain Lowlands, where various agricultural land-use practices have created a diversity of grassland types. Since the 1960s, these agricultural grasslands have diminished as a result of development, farm abandonment, and natural vegetational succession (U.S. Dept. Agric. 1991). This rapid rate of habitat conversion has important implications for the long-term viability of grassland bird populations in Vermont.

This report summarizes data from monitoring programs that focused on or included grassland birds in Vermont. Although sampling generally has been inadequate (i.e., poorly standardized data collection, inconsistent reporting, and/or small sample sizes), these data provide the only means currently available to evaluate the status of Vermont's grassland birds.

Research and Monitoring Projects

Breeding Bird Survey

The U.S. Fish and Wildlife Service's (USFWS) Breeding Bird Survey (BBS) is the best long-term standardized database available for assessing bird population trends in Vermont. Analyses of trend data, provided by the USFWS, suggest an overall decline in many grassland bird populations.

Of 10 species of grassland birds sampled by the BBS in Vermont (Table 1), 8 declined over a 26-year period (1966-1992). However, sample size was large enough for the trend to be statistically significant only for Vesper Sparrow (*Pooecetes gramineus*; Table 1). Over a shorter time period (1982-1991), four of the eight species declined (Table 1), but sample sizes were again too small for the trends to be meaningful for any species. Because of extremely small samples, no trend information was available for Grasshopper Sparrow (*Ammodramus savannarum*) and Henslow's Sparrow (*A. henslowii*) for this time period.

Although these results suggest that many populations of grassland birds in Vermont have declined, they also clearly indicate that sample sizes were generally inadequate to detect meaningful population trends for many species. With only 4 of the 23 BBS routes in Vermont located in the Champlain Lowlands, an expanded sampling effort that specifically targets grassland habitats is required to validate these trends. Although roadside point-counts may not be appropriate for censusing rare species (i.e., Henslow's Sparrow), more common species (i.e., Eastern Meadowlark [*Sturnella magna*] and Bobolink [*Dolichonyx oryzivorus*]) may be adequately sampled with conventional roadside surveys (Vickery 1995).

Records of Vermont Birds

Records of Vermont Birds (RVB), a quarterly publication of the Vermont Institute of Natural Science (VINS), was evaluated as another potential source of information on grassland bird population trends in the state. RVB has published the observations of birdwatchers on a seasonal basis since 1973.

In contrast to the BBS data, results of linear regressions of RVB breeding season totals for each species over time (1973-1992) suggest that most grassland bird populations have increased (Table 2). The only RVB population-trend data that agreed with the BBS data were for Grasshopper Sparrow (Table 2). Nonstandardized reporting and incomplete records potentially confound the reliability of trend information derived from RVB, however. Changes in numbers of

Table 1. U.S. Fish and Wildlife Service Breeding Bird Survey (BBS) trend data for Vermont, 1966–1992 and 1982–1991.

Species	Trends[1] 1966–1992 1982–1991	Sample size[2] 1966–1992 1982–1991	Abundance[3] 1966–1992 1982–1991
American Kestrel	+2.1**[4] -1.4	22 20	0.61 0.63
Upland Sandpiper	-2.4*** -0.6	5 2	0.02 0.16
Horned Lark	-4.5*** +2.3	7 3	0.07 0.13
Sedge Wren	+0.4 +4.8	2 2	0.01 0.05
Vesper Sparrow	-3.8** -21.2**	16 8	0.27 0.20
Savannah Sparrow	-0.7 +7.7	22 20	3.09 3.42
Grasshopper Sparrow	-1.3*** no trend data	5 —	0.04 —
Henslow's Sparrow	-1.0 no trend data	2 —	0.01 —
Bobolink	-1.8 +0.8	24 22	16.38 14.04
Eastern Meadowlark	-3.3 -1.4	23 20	3.09 2.58

[1] Population trends are expressed as the average percent annual change (B. Peterjohn pers. comm.).

[2] Sample size = the number of BBS routes used in the analysis. Trends based on sample sizes smaller than 14 may not be reliable (B. Peterjohn pers. comm.).

[3] Abundance = average number of individuals recorded on routes in the analysis period. For abundances of fewer than 1.0, there may be a positive bias to trend estimates (B. Peterjohn pers. comm.).

[4] Statistical significance levels: * = $0.05 < p < 0.10$; ** = $0.01 < p < 0.05$; *** = $p < 0.01$.

individuals over time most likely reflected increases or decreases in the number of reported sightings. Although RVB is useful in documenting sightings, particularly of rare or unusual birds, we determined it to be of questionable use in assessing population trends.

Upland Sandpiper Survey

Concern over the status of the Upland Sandpiper (*Bartramia longicauda*), a threatened species in Vermont, prompted the initiation

Table 2. Summary of grassland bird population information and nesting status, including trends calculated from Records of Vermont Birds (RVB) reports (1973-1992).

Species	Population estimate[1]	Population trend	Distribution[2]	Nesting status[3]	State status[4]
Northern Harrier	25 ± 5 pairs	BBS: ? RVB: +	localized	S	SC
American Kestrel	unknown	BBS: + RVB: −	widespread	R	—
Upland Sandpiper	100 ± 20 pairs	BBS: − RVB: ?	localized	R	T
Short-eared Owl	confirmed 3X in 1980s	BBS: ? RVB: ?	highly localized	S	SC
Horned Lark	unknown	BBS: − RVB: +	localized	R	—
Sedge Wren	0–2 pairs	BBS: + RVB: ?	highly localized	S	T
Loggerhead Shrike	extirpated	BBS: ? RVB: ?	highly localized	X	E
Vesper Sparrow	unknown	BBS: − RVB: +	localized	R	—
Savannah Sparrow	unknown	BBS: − RVB: +	widespread	R	—
Grasshopper Sparrow	15 ± 5 pairs	BBS: − RVB: −	highly localized	R	SC
Henslow's Sparrow	extirpated	BBS: ? RVB: ?	highly localized	X	E
Bobolink	unknown	BBS: − RVB: +	widespread	R	—
Eastern Meadowlark	unknown	BBS: − RVB: +	widespread	R	—

[1]Represents best-guess estimates from Breeding Bird Surveys (BBS) and from sightings reported in RVB.

[2]Based on information reported in Laughlin and Kibbe 1985
    Localized: occurs in > 5 but < 50% of atlas priority blocks
    Widespread: occurs in > 50% of atlas priority blocks
    Highly localized: occurs in < 5% of atlas priority blocks

[3]Nesting status codes:
    R = Regular nester
    S = Sporadic nester
    X = Extirpated

[4]Vermont status codes (Vermont Dept. Fish Wildl. 1992):
    E = Endangered
    T = Threatened
    SC = Species of Special Concern

of roadside sampling for this species in 1988. From 1988 to 1991, roadside surveys, using taped broadcasts of breeding-season calls (Rimmer and Fichtel 1989), were conducted to determine the presence or absence of Upland Sandpipers in appropriate open grassland habitats.

The number of Upland Sandpipers detected over the four-year period (Table 3) increased from a minimum of 31 in 1988 to a maximum of 128 in 1991 (Rimmer and Fichtel 1989, Peterson et al. 1991). These increases probably reflect increased survey efforts, however, and improved ability to detect the species. In the first year of the survey (1988), only 11 towns were surveyed, whereas 45 towns were surveyed in 1991.

Although Upland Sandpiper numbers appear to be stable in Vermont, no data are available on the species' breeding ecology, productivity, or recruitment. Habitat use and the effects of various land-use practices on this species need to be more rigorously quantified.

Table 3. Summary of results of Upland Sandpiper roadside survey in Vermont.

	1988	1989	1990	1991
Individuals (*n*)	31	79	88	128
Towns Surveyed (*n*)	11	36	24	45

Loggerhead Shrike and Henslow's Sparrow Surveys

The Loggerhead Shrike (*Lanius ludovicianus*) and Henslow's Sparrow, both listed as endangered in Vermont, were surveyed in the state in 1993 and 1992, respectively (Ellison 1992, Peterson and Fichtel 1993). It is believed that these species were never common in Vermont, and nesting has been documented only rarely since the late 1940s. The last known nesting of Loggerhead Shrike in Vermont occurred in 1978 in Panton (Addison Co; Peterson and Fichtel 1993), whereas Henslow's Sparrow was last confirmed nesting in 1953 in Saxtons River (Windham Co; Ellison 1992).

Roadside surveys, in which observers stopped along roadsides in all appropriate habitat and listened for the target species, were conducted for Henslow's Sparrow in 1992 (Ellison 1992) and for Loggerhead Shrike in 1993 (Peterson and Fichtel 1993). For Henslow's Sparrow surveys only, taped broadcasts of songs were used in an attempt to determine the presence of this species. Twenty-eight and 30 towns in the Champlain Lowlands were surveyed for Henslow's Sparrow and

Loggerhead Shrike, respectively. Despite the presence of apparently suitable habitat, no individuals of either species were observed. Both the Loggerhead Shrike and Henslow's Sparrow appear to be extirpated in Vermont.

Atlas of Breeding Birds of Vermont

Vermont's breeding bird atlas project was designed to document the breeding status and distribution of birds throughout the state and is another valuable source of information on grassland birds. Atlas field work was conducted from 1976 to 1981. The state was divided into a grid, and 179 randomly selected priority blocks (each approximately 25 kilometers [km] square) were surveyed to determine the breeding status of all Vermont bird species (Laughlin and Kibbe 1985).

The atlas confirmed the importance of the Champlain Lowlands as the state's most important area for grassland birds. The atlas also documented that grasslands in other parts of the state, such as the Connecticut River valley and Lake Memphremagog, supported diverse grassland bird communities. Although the atlas provided no information on abundance within priority blocks, it did provide some indication of the relative abundance of grassland species (Table 4). A breeding species had to be confirmed as a breeder only once, however, to be confirmed for the entire block. Furthermore, some species may have occurred in fewer blocks than others but may have been very common within those blocks. Additionally, secretive and nocturnal species, such as Short-eared Owl (*Asio flammeus*) and Henslow's Sparrow, were more difficult to confirm and may have been underrepresented in the atlas. For these reasons, caution must be used in applying atlas data to determine relative abundance.

Although the breeding bird atlas has provided valuable information overall, its value would be greatly enhanced by the implementation of a second atlas project. A follow-up breeding bird atlas would yield important information on changes in grassland bird distribution and perhaps relative abundance.

Summary

Currently, little is known about the population status of most grassland bird species in Vermont. The research conducted to date, however, suggests that many grassland species have experienced population declines and that these declines are probably continuing. Two species, Loggerhead Shrike and Henslow's Sparrow, are considered extirpated in the state. Others, such as Short-eared Owl, Sedge Wren (*Cistothorus platensis*), and Grasshopper Sparrow, may be on

Table 4. Breeding status and relative abundance of 13 grassland bird species in Vermont as determined from the Vermont Breeding Bird Atlas.

Species	Number (and percentage) of blocks in which species recorded	Number of blocks in which species recorded as probable breeder[a]	Number of blocks in which species recorded as confirmed breeder[b]
Henslow's Sparrow	1 (0.6%)	0	0
Short-eared Owl	1 (0.6%)	0	1
Sedge Wren	2 (1.0%)	1	0
Loggerhead Shrike	2 (1.0%)	0	1
Grasshopper Sparrow	4 (2.0%)	2	2
Horned Lark	12 (7.0%)	7	3
Upland Sandpiper	13 (7.0%)	5	5
Northern Harrier	22 (12.0%)	3	3
Vesper Sparrow	35 (20.0%)	11	8
Savannah Sparrow	113 (63.0%)	46	38
Eastern Meadowlark	115 (64.0%)	33	56
American Kestrel	140 (78.0%)	17	72
Bobolink	156 (87.0%)	22	125

[a]Probable breeding status = presence of a singing male, a territorial pair, birds exhibiting courtship or agitated behavior, or a bird visiting a probable nest site.
[b]Confirmed breeding status = presence of birds exhibiting distraction displays or nest-building, a female with egg in oviduct, fledged young, adult carrying fecal sac, adult with food for young, or nests with eggs.

the verge of extirpation. More focused and standardized monitoring is needed to determine the status of these and other grassland bird species and to guide conservation efforts.

Recommendations for Future Research

1. Conduct a follow-up statewide breeding bird atlas to analyze changes in grassland bird distribution and relative abundance.
2. Establish roadside and off-road monitoring programs in the Champlain Lowlands and other areas in Vermont to track trends of the more common grassland bird species (i.e., American Kestrel [*Falco sparverius*], Savannah Sparrow [*Passerculus sandwichensis*], Bobolink, and Eastern Meadowlark).
3. Continue or initiate focused surveys for less common and/or apparently extirpated nesting grassland species.
4. Monitor changes in extent and spatial distribution of grasslands and in land-use practices, particularly in the Champlain Lowlands.
5. Expand monitoring efforts to include poorly sampled grasslands in the Connecticut River valley and the Northeast Highlands.
6. Initiate studies of the breeding ecology and population dynamics of grassland species of special concern.
7. Evaluate the effects of various agricultural practices on grassland bird populations, and work with landowners to improve habitats for grassland birds.
8. Develop initiatives to promote economically viable working farms that can serve as reserves for grassland bird species.

Acknowledgments

We extend special thanks to the dedicated volunteers and the staff people of the Vermont Institute of Natural Sciences and Vermont Department of Fish and Wildlife who collected the data summarized in this report. B. Peterjohn of the Breeding Bird Survey and K. Potter of the U.S. Department of Agriculture Soil Conservation Service are also gratefully acknowledged for their contributions of data and other useful information. We are grateful for funding from the Nongame and Natural Heritage Program of the Vermont Department of Fish and Wildlife, The Nature Conservancy, and the U.S. Fish and Wildlife Service, whose support made these projects possible.

Literature Cited

Ellison, W. G. 1992. Vermont status report 1992: Henslow's Sparrow. Vermont Inst. Nat. Sci., Woodstock.

Laughlin, S. B., and D. P. Kibbe, eds. 1985. The atlas of breeding birds of Vermont. Univ. Press of New England, Hanover, NH.

Peterson, J., and C. Fichtel. 1993. Breeding status of the Loggerhead Shrike (*Lanius ludovicianus migrans*) in the Lake Champlain watershed of Vermont. Nongame and Nat. Heritage Prog., Vermont Dept. Fish Wildl., Waterbury.

Peterson, J., C. Rimmer, and C. Fichtel. 1991. A comparison of two population monitoring techniques for Upland Sandpipers in Vermont. Nongame and Natural Heritage Prog., Vermont Dept. Fish Wildl., Waterbury.

Rimmer, C. C., and C. C. Fichtel. 1989. A breeding season status survey of the Upland Sandpiper in Vermont. Tech. Rep. no. 6., Nongame and Natural Heritage Prog., Vermont Dept. Fish Wildl., Waterbury.

U.S. Department of Agriculture. 1991. The changing Vermont landscape II: a resource inventory report. U.S. Dept. Agric. Soil Conserv. Serv. and Vermont Dept. Agric., Winooski, VT.

Vermont Department of Fish and Wildlife. 1992. Endangered and threatened animals of Vermont. Nongame and Natural Heritage Prog., Vermont Dept. Fish Wildl., Waterbury.

Vickery, P. D. 1995. Grassland bird detectability in New England. Center for Biological Conservation, Massachusetts Audubon Soc., Lincoln.

Charles Darmstadt: Vermont Institute of Natural Science, Woodstock, VT 05091.

Christopher Rimmer: Vermont Institute of Natural Science, Woodstock, VT 05091.

Judith Peterson: Vermont Institute of Natural Science, Woodstock, VT 05091.

Christopher Fichtel: The Nature Conservancy, Montpelier, VT 05602.

Grassland Bird Habitat Restoration at Floyd Bennett Field, Brooklyn, New York: Research and Management

Richard A. Lent, Thomas S. Litwin,
Robert P. Cook, Jean Bourque,
Ronald Bourque, and John T. Tanacredi

Floyd Bennett Field (FBF) in Brooklyn, NY, is a 579-hectare (ha) site that provides a significant upland complement to the predominantly estuarine habitats of Jamaica Bay Wildlife Refuge. Historically important as New York City's first airport, FBF is now managed by the National Park Service as part of Gateway National Recreation Area. Since the late 1970s FBF has been recognized as a regionally significant habitat for grassland-dependent birds. Breeding species include Northern Harrier (*Circus cyaneus*), Short-eared Owl (*Asio flammeus*), American Kestrel (*Falco sparverius*), Barn Owl (*Tyto alba*), Upland Sandpiper (*Bartramia longicauda*), Eastern Meadowlark (*Sturnella magna*), and Grasshopper Sparrow (*Ammodramus savannarum*). With successional processes reducing the quantity and quality of grassland habitat, there has been concern over the long-term viability of these species at FBF. This concern is heightened by the fact that most of the above species are officially recognized by New York State as threatened or of special concern, and that FBF is the only protected site of its kind in the New York City area.

We initiated a study of bird-habitat relationships at FBF in 1984. We used the presence of breeding songbirds as indicators of vegetation change. A permanent grid of 60 points, established throughout FBF in 1984 (Fig. 1), was used as a spatial reference for bird and vegetation sampling. Bird populations were surveyed from 1984 through 1987

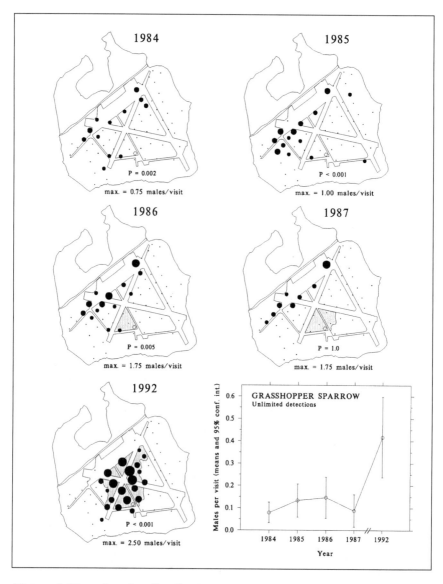

Figure 1. Maps showing distribution and abundance of Grasshopper Sparrows at Floyd Bennett Field, 1984–1992. The FBF boundary, runway system, managed area (shaded), and 60-point sampling grid are shown. Bird abundance at each sampling point is shown by dots, with sizes proportional to mean number of singing males; smallest dot equals zero males, maximum abundance with the largest dot is shown below each map. P-values indicate overall significance of the pattern of abundance. Lower right graph shows a time series of mean abundance, indicating the overall population trend.

and again in 1992 using a series of 10-minute point counts during the peak breeding season, from late May to early July. Vegetation structure was sampled each breeding season in the summers of 1984, 1986, 1987, and 1992 in 0.04-ha plots centered on bird survey points. On a broader scale, we mapped vegetation cover-types using a geographical information system, which allowed us to overlay bird abundance maps onto vegetation maps. Using this database, we conducted a multiscale (macrohabitat, microhabitat, and time) analysis of bird-habitat relationships. Here we emphasize spatial and temporal changes in bird distribution and abundance that parallel changes in habitat availability resulting from active management.

Our initial research, conducted in 1984, identified key grassland areas at FBF where habitat restoration would be most effective. From 1985 through 1990, all woody vegetation was cut or bulldozed from 57 ha of mixed grassland and shrub thicket. Habitat restoration was concentrated in the central infields bounded by the runway system. Follow-up management included cutting resprouts and mowing semiannually (outside the bird-nesting season). The height and extent of woody plant cover were reduced, and the extent of grass and herbaceous cover increased. Previously fragmented grasslands were consolidated into larger, contiguous areas.

Avian responses varied by species. Grasshopper Sparrows, which had been declining at FBF since at least 1980 (R. Bourque and J. Bourque unpubl. data), increased significantly in distribution and abundance between 1984 and 1992 and responded well to habitat management (Fig. 1). Two pairs of Savannah Sparrows (*Passerculus sandwichensis*) nested for the first time in 1989, and the population increased to five pairs in 1992. One pair each of Northern Harriers and Upland Sandpipers nested in 1990 and 1992, respectively. Eastern Meadowlarks, however, declined, from nine pairs in 1982 (R. Bourque and J. Bourque unpubl. data) to none in 1992. Populations of shrub-nesting species, such as Gray Catbirds (*Dumetella carolinensis*; Fig. 2), increased on either side of the central grassland management area.

These patterns of change in avian distribution and abundance strongly suggest that the grassland restoration project at FBF is succeeding in creating new habitat for grassland birds while maintaining populations of shrubland species in other parts of the field. New research initiated in 1993 is addressing the question of whether prescribed fire or mowing is a more effective management technique.

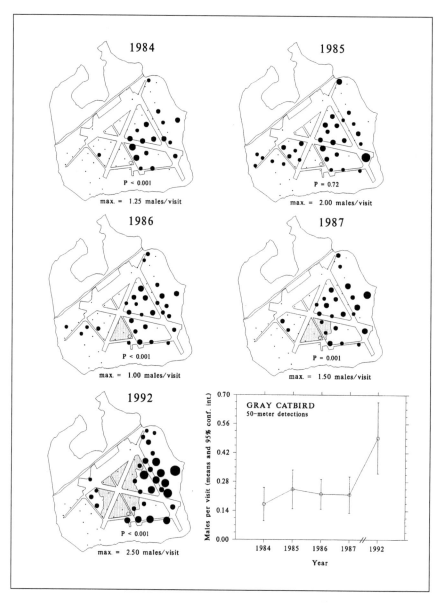

Figure 2. Maps showing distribution and abundance of Gray Catbirds at Floyd Bennett Field, 1984–1992. Map features as in Figure 1.

Acknowledgments

This work was conducted under the auspices of the Seatuck Foundation, Long Island, NY. Funding was provided by the National Park Service, New York City Audubon Society, and Eastern National Parks and Monument Association. N. Giffen (1984), B. Rollfinke (1984), C. Shaw (1985-1987), and P. Champlin (1992) were the field technicians. A. Ducey-Ortiz digitized maps and assisted with data management.

Richard A. Lent: Harvard University, Harvard Forest, P. O. Box 68, Petersham, MA 01366.

Thomas S. Litwin: Clark Science Center, Smith College, Northampton, MA 01063.

Robert P. Cook: National Park Service, Gateway National Recreation Area, Floyd Bennett Field, Brooklyn, NY 11234.

Jean Bourque: New York City Audubon Society, 71 West 23d Street, Suite 1828, New York, NY 10010.

Ronald Bourque: New York City Audubon Society, 71 West 23d Street, Suite 1828, New York, NY 10010.

John T. Tanacredi: National Park Service, Gateway National Recreation Area, Floyd Bennett Field, Brooklyn, NY 11234.

Lepidopteran Assemblages and the Management of Sandplain Communities on Martha's Vineyard, Massachusetts

Paul Z. Goldstein

Introduction

Maritime grasslands, heathlands, and barrens comprise a dynamic mosaic of threatened early-successional natural areas in southern New England. Since the late 1800s, more than 90 percent of the grasslands and heathlands in this region have been lost to development or to the suppression of natural disturbances (Dunwiddie 1994, Noss et al. 1995). The precipitous decline of these communities epitomizes certain dilemmas of managing complex natural areas and presents conservationists with the challenge of maintaining and, where possible, restoring some of southern New England's last remaining early-successional landscapes.

Conservation biologists and entomologists alike have noted the value of insects in conservation programs generally (e.g., Pyle et al. 1981, Collins and Thomas 1991, Kremen et al. 1993), and various authors (e.g., Powell 1978, Opler 1981, Panzer 1983, 1988, Nothnagel et al. 1992, Williams 1993, Panzer et al. 1995) have emphasized the role of insects in evaluating threat, uniqueness, and restoration success in specific habitats or areas. Yet despite a substantial body of literature devoted to the need for invertebrate-based information in managing natural areas, conservation programs in southern New England's sandplain habitats have yet to incorporate such information.

This paper examines the relevance of insect studies to the management of coastal sandplain habitats on Martha's Vineyard, MA, specifically and in southern New England generally. I examine some aspects of species-based information relevant to management at community and ecosystem levels, and I illustrate pitfalls in managing disturbance-prone habitats. At issue are three common misconceptions: (1) that management for a single species or subset of species, particularly those insensitive to environmental change, somehow safeguards a broader natural community; (2) that local species diversity, regardless of threat, should be a management goal; and (3) that ecosystems are static and can be managed as such.

Ecological History of the Martha's Vineyard Sandplain Mosaic

The term "sandplain" refers to a variety of related communities, including pitch pine (*Pinus rigida*)/scrub oak (*Quercus ilicifolia*) barrens and possibly semidegraded oak savanna, in addition to the more popularly known maritime grasslands and heathlands. These habitat "types" are not necessarily mutually exclusive, either spatially, temporally, or in terms of the organisms that comprise them. Traditionally, plant communities have been classified according to species composition, relative abundance, and/or overall vegetative structure. Most sandplain communities, moreover, depend on periodic disturbance (e.g., fire) for their maintenance. Because of this dynamic, these communities are always in a state of flux, with some communities continually grading into and replacing others. Insects do not necessarily distribute themselves according to our definitions of plant communities, however, and our definitions of these communities in botanical terms alone may not be adequate to address the management needs of threatened invertebrates.

As the Wisconsin glaciers retreated northward approximately 12,000 to 16,000 years ago, an expanse of prairie habitats is thought to have occupied a belt across much of the northern United States (Adams 1902, Stuckey 1981). Although their origins were glacial, such areas probably were maintained by a combination of fire and grazing by mammals (Pyne 1982). Although the eastern limit of this "prairie peninsula" continues to be debated (Wright 1968), its eastward extension may have been responsible for the occurrence of certain grassland obligates on Martha's Vineyard. Others may have been adventive, colonizing only after European agricultural practices homogenized much of the landscape. As glacial melting inundated most of the coastal plain, Martha's Vineyard, Long Island, NY, and the New Jersey pine barrens became disjunct remnants, explaining in part the occurrence of several "southern" organisms on Martha's Vineyard.

The southern, western, and post-agricultural ecological affinities of Martha's Vineyard are evident in its invertebrate fauna. Lepidopteran surveys on Martha's Vineyard and on Nantucket Island, MA, documented 110 extralimital species, of which 90 were primarily austral species reaching their recorded known limits on one or both islands, 16 were eastern disjuncts, and 4 were boreal species (Jones and Kimball 1943). A reexamination of these results and of the results of more recent surveys (P. Z. Goldstein unpub. data) reveals that the barrens species are well represented among the southern extralimitals, whereas the eastern disjuncts are often grassland obligates. Several of the grassland-restricted eastern disjuncts such as the cutworm moth *Abagrotis crumbi benjamini*, now listed as a species of special concern in Massachusetts, are prairie species whose nearest extant populations are 3,200 kilometers (km) away. Such organisms may represent relict populations of species that crossed the prairie peninsula.

Using photographs to compare overall community appearance over time, Dunwiddie (1992, 1994) identified broad-scale landscape changes on both Martha's Vineyard and Nantucket since the late 1800s; open habitats declined on both islands. Historically, prairies and grasslands in various parts of North America appear to have been maintained by a combination of fires (including those set by Native Americans; Pyne 1982), grazing by domesticated animals (Howe 1994), and possibly weather-related disturbances such as storms and salt spray near the coast (P. W. Dunwiddie pers. comm.). On Martha's Vineyard, grazing was once so extensive that a Revolutionary War account estimated that more than 10,000 sheep were seized by British soldiers in a single raid in 1778 (Hoyt 1978). With the decline of agriculture and of widespread grazing on the island beginning around the 1880s, and with the development of fire-suppression techniques, the processes for maintaining early-successional habitats were interrupted, explaining in part why these habitats changed so drastically (Dunwiddie 1994).

As early-successional communities have declined on Martha's Vineyard during the twentieth century, so too has the lepidopteran fauna associated with them. One of the best known examples of a regionally extirpated insect is the regal fritillary (*Speyeria idalia*). Described as "widely distributed and common; in especially large numbers at flowers of the swamp milkweed" (Jones and Kimball 1943), this butterfly remained abundant throughout much of the island's outwash plain through most of the 1960s (D. C. Ferguson pers. comm.). As of the mid-1980s there were five known sites in New England where the butterfly remained, all on grassland habitats on offshore islands (Martha's Vineyard, Nantucket, Naushon Island [MA],

No Mans Land [MA], and Block Island [RI]; Mello 1989, Schweitzer 1992). Beginning with Martha's Vineyard, populations on each of the islands crashed. The butterfly was last reported from Martha's Vineyard in 1986 (P. Goldstein unpubl. data), from Naushon in 1988 (Mello 1988 and pers. comm.), from Nantucket in 1990 (P. W. Dunwiddie pers. comm.), and from Block Island in 1991 (L. Gall pers. comm.). Searches in 1992 failed to locate individuals on any of these islands (Goldstein 1992a, Wagner et al. 1997).

The regal fritillary is not the only butterfly that has declined on Martha's Vineyard. Jones and Kimball (1943) reported five species of fritillaries on the island. Although two of these (Aphrodite fritillary [*Speyeria aphrodite*] and meadow fritillary [*Boloria bellona*]) are known only from single records, I have not recorded the other three (except the regal fritillary prior to its disappearance) in more than 20 years of collecting on the island. Jones and Kimball (1943) were puzzled by the rarity of the great spangled fritillary (*S. cybele*), describing it as "unaccountably rare, though the traditional food plants of its larvae (violets) abound." Several moths also appear to have declined on the island (P. Z. Goldstein unpubl. data). Grassland-obligate and grass-feeding species are over-represented among the species that have declined, paralleling the decline of grasslands during the twentieth century. A list of new lepidopteran records from the island, excluding single records and strays, is dominated by tree-foliage-feeding species. The only new record of a grass-feeding moth is the polyphagous introduced European pest species of cutworm *Noctua pronuba*.

Since 1986 more than 25 species of rare and endangered lepidopterans have been documented on the sandplains of Martha's Vineyard (see Appendix). Two of these species (a tiger moth [*Cycnia inopinatus*] and the woolly gray geometrid [*Lycia ypsilon*]) are known nowhere else in New England and appear to reach their northern limits on Martha's Vineyard. Another three species (imperial moth [*Eacles imperialis*], the noctuid *Apamea mixta*, and barren's Metarranthis [*Metarranthis apiciaria*]) are believed to be extirpated or absent from mainland New England. Four species (Melsheimer's sack-bearer [*Cicinnus melsheimeri*] and the noctuids *Abagrotis crumbi benjamini, Cucullia speyeri*, and decodon stemborer [*Papaipema sulphurata*]) are represented in New England only by populations on Cape Cod and the offshore islands of Massachusetts. The decodon stemborer is endemic to southern coastal New England (Bird 1926), and the Martha's Vineyard population of *M. apiciaria* is the world's only known extant population of this species. A majority of these rare organisms, including the imperial moth, woolly gray, *M. apiciaria*, and Melsheimer's sack-bearer, are associated exclusively with pitch pine/scrub oak barrens.

Ecosystem Management and Species Diversity

The complexity of the Martha's Vineyard sandplain demands that its varied components not be managed in isolation, and efforts are underway to pursue a broad approach to managing the island's conservation areas. Several questions confront those who will implement that management, however. The most fundamental is, what are the specific goals of the management program with respect to each of the components of the sandplain? To maintain the known assemblage of rare organisms? To maintain a select few of those species under the assumption that the rest will follow suit? Or perhaps to maintain or restore a particular habitat "architecture" believed to have been more widespread historically?

The understanding that it is not feasible to manage natural areas on a per-species basis has led to the popular paradigm of "ecosystem management," which aims to protect not only the elements of biological diversity but the processes that spawn and regenerate those elements. It should be emphasized, however, that the mechanics of such management processes are essentially unknowable, and that axiomatic definitions of "ecosystem" allow one to justify virtually any protocol as "management" (Wilcove and Blair 1995). In the case of the remaining sandplains on Martha's Vineyard, it is the application of plant-community definitions to broader management goals that may not be justified. This is a severe drawback to the application of what may amount to arbitrary management practices. Ecosystem management can be successful only when "ecosystem" is defined with precision and when the criteria used actually reflect the needs of the natural entities requiring protection.

When asked what the relevance of insects is to ecosystem management, most biologists are likely to cite species diversity as fundamentally important. Arthropods comprise the most species-rich animal phylum. Not only is the profound richness of insect species (estimated to be between 10 and 30 million taxa; Erwin 1988, Stork 1988, Wilson 1988) impressive, but relative to other animals, insects and other arthropods tend to dominate most terrestrial habitats at local as well as at regional and global scales in terms of species richness, biomass, and abundance (Erwin 1982, 1988, Stork 1988, Wilson 1988, Kremen et al. 1993). Species diversity is not necessarily related to an area's uniqueness or vulnerability, however. The relevance of insects lies not in their species richness but in their specialization and, relatedly, their sensitivity to environmental change. It is by virtue of that sensitivity that insects may add a new dimension in which to evaluate management.

To the extent that conservation aims to protect unique and vulnerable areas, local species richness might well be decoupled from conservation priority (Vane-Wright et al. 1991, Nixon and Wheeler 1992). Simple measures, indeed any measures, of diversity are not relevant to stewardship if much or all of the diversity consists of species that are exotic, for example (Noss 1990), or if the analysis of monitoring data lumps species in ways that obscure the effects of management on species that are of concern. In southern New England, critically imperiled habitats, including barrier beaches and sandplain grasslands, may be rather species poor. But does this obviate attempts to conserve these areas, given that their species comprise an assemblage of rare or threatened habitat-obligates? Even if one chooses to manage species diversity at local scales, there is no predictable relationship between species diversity in unrelated groups (Goldstein in press).

Insects provide a special challenge when evaluating management practices. In some cases, researchers have lumped individuals of different species such that the effects of the management practice on those species are masked. In several studies on the effects of fire on invertebrates, for example, individuals were lumped across species or higher taxa (e.g., Cancelado and Yonke 1970, Seastedt 1984a, b, Dunwiddie 1991). This approach assumes that sets of differentially related organisms are somehow equivalent biologically (e.g., with respect to ecology and natural history) yet independent statistically. Simply counting numbers of individuals belonging to a given higher taxon to assess the effects of a particular treatment on "densities" or "abundance" ignores the relative abundance of species within the sample, thus assuming equal distributions of species (species evenness) and conflating diversity with abundance (see Hurlbert 1971). A fourfold increase in "abundance" or "density" within a sample of wasps is hardly meaningful to a land manager if the increase can be explained by a 500 percent increase in the numbers of one exotic species accompanied by the extirpation of 10 native habitat obligates. Sampling effects hardly need to be as exaggerated as this example to affect the results of such an analysis in a profound way.

I am not suggesting that land managers look at populations of every species. Managers should study as many species as feasible, but they should analyze the effects of treatments (including fire) on those species separately in order to identify conflicts in how the different organisms respond to treatments. An understanding of the most sensitive or vulnerable organisms not only enhances our ability to identify the most critically threatened natural community types, but also provides a conservative estimate of the thresholds of management risk.

Sensitivity of Insects to Environmental Change

Insects, particularly herbivorous insects such as most Lepidopera, are extremely sensitive gauges of environmental change (Collins and Thomas 1991, Forey et al. 1994, Samways 1994). Arthropods are generally susceptible to a variety of changes, including climatic, atmospheric, and biotic (Fajer 1989, Fajer et al. 1989, Erhardt and Thomas 1991, Kremen et al. 1993). A partial explanation for their sensitivity may be found in a biological attribute common to many insects and other arthropods: the feature of specialization, which several authors have implicated in susceptibility to extinction (Fowler and McMahon 1982, Panzer 1988, Wagner and Liebherr 1992, Wiegmann et al. 1993). Obligate year-round association with a particular habitat is one manifestation of specialization, which may be mediated by nutritional requirements, for example. Herbivorous insects are often notoriously oligophagous, with a majority feeding on one or a few closely related plants (Futuyma 1991, Mitter and Farrell 1991, Farrell and Mitter 1993). And specialization along any axis is compounded in holometabolous insects (those that undergo complete metamorphosis), whose larvae, adults, and even pupae may have nonoverlapping—but equally specific—requirements. Other features that may be related to vulnerability (particularly among prairie and grassland species) include fluctuating population densities and generally poor dispersal abilities (Panzer 1988), although the latter makes little intuitive sense for organisms living in disturbance-prone areas. Most importantly, insects appear to have requirements that are simply not well understood.

Butterflies and moths epitomize the sensitivity of arthropods to environmental change, as well as our poor understanding of their vulnerability. On Martha's Vineyard and Nantucket Island, where fritillary butterflies have declined dramatically, the fritillaries' larval food plants (violets [*Viola* spp.]) appear to remain abundant enough to support populations of herbivores. But fritillaries appear to be vulnerable to extinction for several reasons, occupying a disproportionately large subset of endangered and threatened butterfly taxa on endangered-species lists in Europe and North America (D. L. Wagner pers. comm.). As many as five species of fritillaries may have been extirpated from Martha's Vineyard since the early 1950s. To date, explanations for their disappearance range from a lowered availability of adult nectar sources to the decline of grazing, oceanic salt spray brought on by hurricanes, and the presence of introduced pathogens.

Insect sensitivity to environmental alteration is particularly obvious when examining faunal change. Erhardt and Thomas (1991) documented widespread declines in the butterfly fauna associated with European grasslands following the industrialization of agricultural practices. They also noted the disappearance of several butterfly species from areas of Europe long before declines in their host plants could be documented, suggesting that these organisms have requirements other than nutritional ones (Erhardt and Thomas 1991). These authors further noted that many lepidopteran larvae have temperature requirements conferred by the architecture of their host plants. In one example, the butterfly *Lysandra bellargus* is "restricted, in Britain, to south-facing slopes in the southern counties, but the 1–2 cm tall *Hippocrepis comosa* plants used by its larvae are about 4–7° warmer than those growing in a 7 cm-tall sward during the main period of spring feeding" (Erhardt and Thomas 1991: 236).

The narrow thermoregulatory requirements of lepidopterans can also be seen in the moth associates of frost pockets in New England barrens. On Martha's Vineyard, frost bottoms are channels that were etched by the southward flow of meltwaters during the retreat of the Wisconsin glaciers (but see Uchupi and Oldale 1994). The northern extensions of the Martha's Vineyard frost bottoms are extremely xeric; southward they grade into the heads of one or more coves of the outwash-plain ponds along the island's south shore. Most of the drier portions of the bottoms are dominated primarily by scrub oak and a variety of other barrens plants with a significant component of grassy cover. These areas undergo extreme temperature fluctuations; by night they are the coldest areas on the island, and by day the hottest. They appear to collect cold air because of downslope nighttime winds, and it is not unusual for frost-bottom temperatures to fluctuate as much as 39 degrees C in a single 24-hour period or to experience frost 10 to 12 months of the year (P. Z. Goldstein unpubl. data). As a result of the frequent frosts, much of the vegetation in these areas (particularly the scrub oak) undergoes notable delays in leaf-out (Aizen and Patterson 1995); scrub oaks in these areas rarely leaf out before June, may be delayed until July, and in some cases may fail to produce mature foliage for an entire year due to late-season frost kill (T. Chase pers. comm.).

Yet these seemingly forbidding conditions may explain why frost bottoms appear to be local refugia for so many rare moths (on Martha's Vineyard, frost bottoms account for the majority of the rare lepidopteran occurrences and are being managed in conjunction with grassland communities; Goldstein 1992b, 1994) and why, for example, maintaining upland patches of host plants will not sustain these animals. Aizen and Patterson (1995) showed empirically that the leaf damage by scrub-oak-feeding thrips (Thysanoptera) in similar frost pockets on Cape Cod is consistent with the "phenological window" hypothesis of Feeny (1970). Feeny suggested that nutritional suitability of certain oaks to lepidopteran caterpillars may change drastically during the growing season, such that herbivores might have only a narrow window of time in which to develop. Noting that hatchling caterpillars of many oak-feeding lepidopterans require young nutrient- and nitrogen-rich foliage (Feeny 1970, Aizen and Patterson 1995), and that many of the apparent frost-bottom obligate species are summer flyers at or near the northern limits of their ranges, it is conceivable that by the time the larvae of some of these species are ready to hatch that most of the upland scrub oak may already be mature and therefore nutritionally unsuitable. Thus the temperature-mediated late leaf-out of frost-bottom scrub oaks might in turn mediate larval nutritional requirement and adult phenology, allowing the moths to persist in these depressions.

An example that appears to reflect a more general pattern of poorly understood resource requirements among herbivorous insects may be found in the specialization on a nutritionally suboptimal host plant by the imperial moth. New England's last extant population of this moth persists on Martha's Vineyard, having been extirpated from mainland New England for reasons that are not well understood but that may have involved pesticides (Goldstein 1991). Laboratory and field experiments indicate that imperial moth larvae grow more quickly, achieve greater body mass, and exhibit higher survivorship when fed post oak (*Quercus stellata*) rather than pitch pine, despite the fact that northern imperial moths (including the Martha's Vineyard population) feed exclusively on pine in the wild (Goldstein 1991 and unpubl. data). These data appear to corroborate what seems to be a general pattern of narrower host preference in ovipositing adult lepidopterans than in their larvae (Wiklund 1975, Chew 1977, Courtney 1981).

Disturbance and Habitat Heterogeneity

As mentioned earlier, sandplain communities form a mosaic in time and space, and this mosaic depends directly on disturbance. Not only does the "structure" of such habitats depend on some form of periodic disturbance, such as fire, but many of the plant and animal species characteristic of such areas are restricted to particular seral stages. Obligate "pioneer species" may have a variety of requirements related to the conditions of early-successional communities, and they are easily crowded out by succession. To the extent that we understand them, species-specific requirements dictate—or should dictate—management goals. Since we do not know all the parameters along which insects are specialized, we must maintain as many habitat types and seral stages as possible.

In nature, the kinds of disturbances that maintain early-successional communities are stochastic, varying unpredictably in magnitude and frequency. No matter what their effects on individual organisms are, disturbances create new habitat for colonization, assuming source colonies exist nearby. In this light, the interruption of disturbance regimes may be viewed as a deterministic threat to early seral communities, posing the most immediate threat to the earliest seral stages.

But uninterrupted ecological succession is only one form of change that threatens habitat heterogeneity. Both natural deterministic threats (Petraitis et al. 1989) and anthropogenic threats, including traditionalized management strategies (Howe 1994), can have a homogenizing effect on natural communities. The early-successional character of sandplains, and the dynamic nature of their components, mandate that no single seral stage can be maintained in perpetuity in a given area. For this and other reasons, most managed areas on Martha's Vineyard that are routinely subjected to prescribed burning, for example, are divided into units to avoid burning too much in any given year. But breaking up areas into burn units is in itself not enough to safeguard heterogeneity. Given that the effects of fire on plant communities vary with the season and with the fire's size, season, and intensity, failure to vary any of these factors within a given treatment area will likely fail to protect much of a range of rare species, and may result in monoculture (Howe 1994).

Maintenance Versus Restoration?

An apparent temptation facing some conservationists is to prioritize components of the sandplain mosaic and to manage certain components in isolation. Thus the term "sandplain" has been used erroneously as a synonym for "sandplain grassland." The recognition that sandplain grasslands and heathlands have lost a higher percentage of area than other sandplain components is indeed cause for concern. However, I submit that the less degraded, more expansive, and more intact habitats such as pine barrens merit at least as much, if not more, attention.

Most sandplain habitats depend to some extent on burning. Some habitats, such as the earliest seral grassland and heathland stages, may depend on more frequent burning than others. Indeed, fire suppression has threatened barrens on Martha's Vineyard and elsewhere only slightly more slowly than it has grasslands, and the encroachment of spruces (*Picea*) and tree oaks on many barrens areas is apparent.

Without an understanding of frost bottoms and their apparent importance as refugia for rare insects, one might opt to "restore" grasslands in such areas, although there is no evidence they ever existed there. Many organisms, including numerous moths as well as the Northern Harrier (*Circus cyaneus*), that were traditionally thought of as grassland/heathland obligates appear to thrive in grassy shrublands that characterize frost bottoms. Highlighting the limited value of plant community definitions, the pine-oak barrens on Martha's Vineyard appear to retain more "grassland" moth species than Katama Plain, the island's most intensively managed sandplain grassland. And whereas most of the rarest and most threatened moths on Martha's Vineyard are associated exclusively with barrens, all but four putative grassland/heathland species (tiger moth, chain-dotted geometer [*Cingilia catenaria*], slender clearwing [*Hemaris gracilis*], and wild cherry sphinx [*Sphinx drupiferarum*]) have been documented from numerous frost bottoms (see Appendix). It is clear that many species currently prioritized in conservation programs based on their association with "pure grasslands" survive and thrive in grassy shrublands. Moreover, it is likely that the abundances of other species of current concern during widespread agriculture were artifacts of that agriculture, and one should not discount the possibility that some of what are now considered threatened grassland species were merely adventive features of an artificial landscape.

This is not to say that barrens should never be burned, that the demise of grasslands should be accepted apathetically, or that managers should feel paralyzed by the lack of available scientific information. The threat of succession and the distribution and habitat associations of insects, combined with our growing understanding of the effects of management practices on habitat structure and function, argue neither for a panicked rush of restoration nor for a laissez-faire approach. On balance, the available information suggests a rigorous approach to management science and an intensive but carefully considered strategy in management practice.

Summary

1. Components of the sandplain mosaic on Martha's Vineyard include maritime grasslands, heathlands, and barrens. The grassland/heathland complex has declined precipitously, with concomitant declines in the lepidopteran fauna. However, the Martha's Vineyard sandplain continues to support a varied assemblage of rare and endangered insects.
2. Barrens, including scrub oak–dominated frost bottoms, are currently more expansive on Martha's Vineyard than grassland/heathland complexes and support most of the known rare and threatened insect species, as well as several species traditionally thought of as grassland obligates.
3. Traditional definitions of plant communities may be inadequate to address the management needs of many rare and endangered organisms.
4. Organisms that display specialization to a variety of conditions are useful in identifying otherwise cryptic community parameters and seral stages. The more conditions that are taken into account when managing an area, the finer the scale at which management should be practiced.
5. Insects such as butterflies and moths are sensitive to environmental alteration and are useful gauges of the success of management and restoration efforts. Their narrow yet varied life history requirements make them especially suitable for helping design the scales at which management practices should be applied.
6. Based on the lepidopteran fauna, maintenance of pitch pine/scrub oak barrens, frost bottoms, and grassy scrublands should receive a higher priority than large-scale restoration of pure grasslands on Martha's Vineyard. In no case should barrens communities be sacrificed to recreate grasslands where the latter may never have existed.

Acknowledgments

T. Simmons, T. Chase, and A. Stevens provided valuable discussions on sandplain invertebrates, fire, and ecosystem management. T. Simmons and K. Turner of the Massachusetts Field Office of The Nature Conservancy were invaluable sources of references, particularly on prescribed burning. A. Brower, R. Colwell, and A. Stevens also provided several helpful references. L. Johnson, J. Varkonda, and E. Littlefield contributed many hours of temperature data from the Martha's Vineyard frost bottoms. B. O'Neill and N. Tutco of the Vineyard Conservation Society also provided valuable assistance. I am indebted to J. Carpenter, T. Simmons, and E. O. Wilson for providing encouragement during the early years of insect sampling on Martha's Vineyard. I thank the National Science Foundation, Environmental Protection Agency, American Museum of Natural History, and Center for Conservation and Biodiversity (= Department of Ecology and Evolutionary Biology, University of Connecticut) for support while writing this paper.

Literature Cited

Adams, C. C. 1902. Postglacial origin and migrations of the life of the northeastern United States. J. Geog. 1: 352-357.

Aizen, M. A., and W. A. Patterson III. 1995. Leaf phenology and herbivory along a temperature gradient: a spatial test of the phenological window hypothesis. J. Veg. Sci. 6: 543-550.

Bird, H. 1926. New life histories and notes in *Papaipema* no. 24 (Lepidoptera). Can. Ent. 57: 249-284.

Cancelado, R., and T. R. Yonke. 1970. Effect of burning on insect populations. J. Kansas Ent. Soc. 43: 274-281.

Chew, F. S. 1977. Coevolution of pierid butterflies and their cruciferous food plants. II. The distribution of eggs on potential food plants. Evolution 31: 568-579.

Collins, N. M., and J. A. Thomas. 1991. The conservation of insects and their habitats. Academic Press, London.

Courtney, S. P. 1981. Coevolution of pierid butterflies and their cruciferous foodplants. III. Survival, development, and oviposition on different plants. Oecologia 51: 91-96.

Dunwiddie, P. W. 1991. Comparisons of aboveground arthropods in burned, mowed, and untreated sites in sandplain grasslands on Nantucket Island, Massachusetts, U.S.A. Am. Midl. Nat. 125(2): 206-212.

Dunwiddie, P. W. 1992. Changing landscapes: a pictorial field guide to a century of change on Nantucket. Nantucket Conserv. Foundation, Nantucket Historical Soc., and Massachusetts Audubon Soc., Nantucket, MA.

Dunwiddie, P. W. 1994. Martha's Vineyard landscapes: the nature of change. Vineyard Conserv. Soc., Vineyard Haven, MA.

Erhardt, A., and J. A. Thomas. 1991. Lepidoptera as indicators of change in the seminatural grasslands of lowland and upland Europe. Pp. 213-236 *in* The conservation of insects and their habitats (N. M. Collins and J. A. Thomas, eds.). Academic Press, London.

Erwin, T. L. 1982. Tropical forests: their richness in Coleoptera and other arthropod species. Coleop. Bull. 36: 74-75.

Erwin, T. L. 1988. The tropical forest canopy: the heart of biotic diversity. Pp. 123-129 *in* Biodiversity (E. O. Wilson, ed.). Natl. Academy Press, Washington, D.C.

Fajer, E. D. 1989. How enriched carbon dioxide environments may alter biotic systems even in the absence of climatic changes. Conserv. Biol. 3: 318-320.

Fajer, E. D., M. D. Bowers, and F. A. Bazzazz. 1989. The effects of enriched carbon dioxide atmospheres on plant-insect herbivore interactions. Science 243: 1198-1200.

Farrell, B. D., and C. Mitter. 1993. Phylogenetic determinants of insect/plant community diversity. Pp. 253-266 *in* Species diversity in ecological communities (R. E. Ricklefs and D. Schluter, eds.). Univ. of Chicago Press, Chicago.

Feeny, P. P. 1970. Seasonal changes in oak leaf tannins as a cause of spring feeding by winter moth caterpillar. Ecology 51: 565-581.

Forbes, W. T. M. 1923. The Lepidoptera of New York and neighboring states. Pt. 1. Cornell Univ. Agric. Exper. Stn. Mem. 68, Ithaca, NY.

Forbes, W. T. M. 1948. The Lepidoptera of New York and neighboring states. Pt. 2. Cornell Univ. Agric. Exper. Stn. Mem. 274, Ithaca, NY.

Forbes, W. T. M. 1954. The Lepidoptera of New York and neighboring states. Pt. 3. Cornell Univ. Agric. Exper. Stn. Mem. 329, Ithaca, NY.

Forey, P. L., C. J. Humphries, and R. I. Vane-Wright, eds. 1994. Systematics and conservation evaluation. Clarendon Press, Oxford, U.K.

Fowler, C. W., and J. A. McMahon. 1982. Selective extinction and speciation: their influence on the structure and functioning of communities and ecosystems. Am. Nat. 119: 480-497.

Futuyma, D. J. 1991. Evolution of host specificity in herbivorous insects: genetic, ecological, and phylogenetic aspects. Pp. 431-454 *in* Plant-animal interactions: evolutionary ecology in tropical and temperate regions (P. W. Price, T. M. Lewinsohn, G. W. Fernandez, and W. W. Benson, eds.). John Wiley, New York.

Goldstein, P. Z. 1991. Natural history and nutritional ecology of a remnant population of *Eacles imperialis* Drury on Martha's Vineyard Island, Massachusetts. Undergrad. thesis, Harvard Univ., Cambridge, MA.

Goldstein, P. Z. 1992a. The final countdown: surveys of offshore islands fail to document *Speyeria idalia*. Report to Massachusetts Natural Heritage and Endangered Species Prog., Boston.

Goldstein, P. Z. 1992b. Informative occurrences of insects in the Manuel Correllus state forest, Martha's Vineyard Island, Massachusetts: comments and management recommendations. Report to Massachusetts Natural Heritage and Endangered Species Prog., Boston.

Goldstein, P. Z. 1994. Update of rare insect occurrences in the Manuel Correllus state forest and vicinity, Martha's Vineyard, Dukes County, Massachusetts. Report to Massachusetts Natural Heritage and Endangered Species Prog., Boston.

Goldstein, P. Z. In press. How many things are there? A reply to Oliver and Beattie. Conserv. Biol.

Howe, H. F. 1994. Managing species diversity in tallgrass prairie: assumptions and implications. Conserv. Biol. 8(3): 691-704.

Hoyt, E. P. 1978. Nantucket: the life of an island. Stephen Greene Press, Brattleboro, VT.

Hurlbert, S. H. 1971. The nonconcept of species diversity: a critique and alternative parameters. Ecology 52(4): 577-586.

Jones, F. M., and C. P. Kimball. 1943. The Lepidoptera of Nantucket and Martha's Vineyard. Nantucket Maria Mitchell Assoc., Nantucket, MA.

Kremen, C., R. K. Colwell, T. L. Erwin, D. D. Murphy, R. F. Noss, and M. A. Sanjayan. 1993. Terrestrial arthropod assemblages: their use in conservation planning. Conserv. Biol. 7(4): 796-808.

Mello, M. 1988. Survey of rare insects on the Elizabeth Islands. Report to Massachusetts Natural Heritage and Endangered Species Prog., Boston.

Mello, M. 1989. Survey and ecology of *Speyeria idalia* (regal fritillary) on Block Island, with notes on other species. Report to The Nature Conservancy, Providence, RI.

Mitter, C., and B. D. Farrell. 1991. Macroevolutionary aspects of insect-plant relationships. Pp. 35-78 *in* Insect plant interactions, vol. 3 (E. Bernays, ed.). CRC Press, Boca Raton, FL.

Nixon, K. C., and Q. D. Wheeler. 1992. Measures of phylogenetic diversity. Pp. 216-234 *in* Extinction and phylogeny (M. J. Novacek and Q. D. Wheeler, eds.). Columbia Univ. Press, New York.

Noss, R. F. 1990. Can we maintain biological and ecological integrity? Conserv. Biol. 4(3): 241-243.

Noss, R. F., E. T. LaRoe III, and J. M. Scott. 1995. Endangered ecosystems of the United States: a preliminary assessment of loss and degradation. Biol. Rep. 28, U.S. Dept. Interior, Natl. Biol. Serv., Washington, D.C.

Nothnagel, P., T. Simmons, and P. Z. Goldstein. 1992. *Cicindela dorsalis dorsalis*. Population and habitat conditions on Martha's Vineyard, 1990-1992. Report to Massachusetts Natural Heritage and Endangered Species Prog., Boston.

Opler, P. A. 1981. Management of prairie habitats for insect conservation. Nat. Areas J. 1(4): 3-6.

Panzer, R. 1983. Fire, insects studied on a prairie hill remnant (Illinois). Restor. & Man. Notes 1(4): 17-18.

Panzer, R. 1988. Managing prairie remnants for insect conservation. Nat. Areas J. 8(2): 83-90.

Panzer, R., D. Stillwaugh, R. Gnaedinger, and G. Derkovitz. 1995. Prevalence of remnant dependence among the prairie- and savanna-inhabiting insects of the Chicago region. Nat. Areas J. 15(2): 101-116.

Petraitis, P. S., R. E. Latham, and R. A. Niesenbaum. 1989. The maintenance of species diversity by disturbance. Quart. Rev. Biol. 64(4): 393-418.

Powell, J. A. 1978. Endangered habitats for insects: California coastal sand dunes. Atala 6: 1-2.

Pyle, R. M., M. Bentzien, and P. Opler. 1981. Insect conservation. Ann. Rev. Ent. 26: 233-258.

Pyne, S. J. 1982. Fire in America: a cultural history of wildland and rural fire. Princeton Univ. Press, Princeton, NJ.

Ross, G. N. 1995. Butterfly wrangling in Louisiana. Nat. Hist. May 1995: 36-43.

Samways, M. J. 1994. Insect conservation biology. Chapman and Hall, New York.

Schweitzer, D. F. 1983. Rare lepidoptera of the Katama Plain on Martha's Vineyard, Massachusetts. Report to Massachusetts Natural Heritage and Endangered Species Prog., Boston.

Schweitzer, D. F. 1992. *Speyeria idalia*, the regal fritillary butterfly: results of a global status survey. Report to U.S. Fish Wildl. Serv., region 5 [no city or state cited].

Seastedt, T. R. 1984a. Belowground microarthropods of annually burned and unburned tallgrass prairie. Am. Midl. Nat. 11(2): 405-408.

Seastedt, T. R. 1984b. Microarthropods of burned and unburned tallgrass prairie. J. Kansas Ent. Soc. 57(3): 468-476.

Stork, N. E. 1988. Insect diversity: facts, fiction and speculation. Biol. J. Linn. Soc. 35: 321-327.

Stuckey, R. L. 1981. Origin and development of the concept of the prairie peninsula. Pp. 4-23 *in* The prairie peninsula—in the shadow of Transeau (R. L. Stuckey and K. J. Reese, eds.). Ohio Biol. Surv. Notes no. 15.

Uchupi, E., and R. N. Oldale. 1994. Spring sapping of the enigmatic relict valleys of Cape Cod and Martha's Vineyard and Nantucket islands, Massachusetts. Geomorphology 9: 83-95.

Vane-Wright, R. I., C. J. Humphries, and P. H. Williams. 1991. What to protect? Systematics and the agony of choice. Biol. Conserv. 55: 235-254.

Wagner, D. L., and J. K. Liebherr. 1992. Flightlessness in insects. Trends Ecol. Evol. 7: 216-220.

Wagner, D. L., M. S. Wallace, J. Boettner, and J. Elkinton. 1997. Status update and life history studies on the regal fritillary (Lepidoptera: Nymphalidae). Pp. 261-276 *in* Grasslands of northeastern North America: ecology and conservation of native and agricultural landscapes (P. D. Vickery and P. W. Dunwiddie, eds.). Massachusetts Audubon Soc., Lincoln.

Wiegmann, B. M., C. Mitter, and B. Farrell. 1993. Diversification of carnivorous parasitic insects: extraordinary radiation or specialized dead end? Am. Nat. 5: 737-754.

Wiklund, C. 1975. The evolutionary relationships between adult oviposition preference and larval host plant range in *Papilio machaon* L. Oecologia 18: 185-197.

Wilcove, D. S., and R. B. Blair. 1995. The ecosystem management bandwagon. Trends Ecol. Evol. 10(8): 345.

Williams, K. S. 1993. Use of terrestrial arthropods to evaluate restored riparian woodlands. Restoration Ecol. June 1993: 107-116.

Wilson, E. O. 1988. The current state of biological diversity. Pp. 3-17 *in* Biodiversity (E. O. Wilson, ed.). Natl. Academy Press, Washington, D.C.

Wright, H. E., Jr. 1968. History of the prairie peninsula. Pp. 78-88 *in* The Quaternary of Illinois (R. E. Bergstrom, ed.). Univ. Ill. Agric. Spec. Publ. 14.

Paul Z. Goldstein: Department of Entomology, American Museum of Natural History, Central Park West at 79th Street, New York, NY 10024.

Appendix. *Occurrences of statewide and regionally rare moths and their primary habitat associations on Martha's Vineyard and Nantucket I., MA. Except where noted, species have been verified from Martha's Vineyard since 1986; historical records are those of Jones and Kimball (1943). Unless otherwise noted, recent documentations of species on Martha's Vineyard are by Goldstein (1992, 1996, unpubl. data).† = listed or proposed for listing as endangered, threatened, or special concern species in Massachusetts; N = Nantucket; MV = Martha's Vineyard; SH = state historical (not documented from Massachusetts since 1950); NA = not applicable. Sources cited may refer to distribution, habitat, or host plant information; DFS = D. F. Schweitzer pers. comm.*

Putative Primary Habitat Associations	Host Plant(s)	Historical Occurrences	Sources	Comments
Barrens/Frost Bottoms				
Arctiidae				
Cisthene packardii	Lichens	None		Occurs in Myles Standish State Forest, Plymouth County
Geometridae				
Itame sp.†	Unknown; presumed *Vaccinium*	NA	DFS	
Lycia ypsilon†	*Clethra, Prunus; Amelanchier* suspected	None	DFS	Southern species; MV (where it occurs only in frost bottoms) is only New England occurrence
Metarranthis apiciaria†	Unknown	None	Forbes 1948	MV is only verified occurrence since 1974; occurs only in frost bottoms
Mimallonidae				
Cicinnus melsheimeri†	*Quercus ilicifolia*	N, MV		Southern species; frost-bottom obligate on MV
Noctuidae				
Acronicta albarufa†	*Prunus*	MV	Forbes 1954	Candidate for federal listing
Apharetra purpurea†	*Vaccinium*	MV	Forbes 1954	Verified on MV by T. Simmons (pers. comm. 1987); probably confused with *A. dentata* by Jones and Kimball (1943)
Catocala herodias gerhardi†	*Quercus ilicifolia*	MV	DFS	

Putative Primary Habitat Associations	Host Plant(s)	Historical Occurrences	Sources	Comments
Barrens/Frost Bottoms *(continued)*				
Noctuidae *(continued)*				
Catocala sp.	*Quercus* spp.	NA	DFS	
Chaetaglaea cerata	*Quercus ilicifolia*	N, MV	Forbes 1954	
Chaetaglaea tremula	*Vaccinium, Comptonia, Quercus*	None	DFS	
Chytonix sensilis	Fungi on rotten wood after fires	MV	Forbes 1954, DFS	May be obligate of burned barrens
Eucoptocnemis fimbriaris	Unknown	N, MV	Forbes 1954	
Lithophane thaxteri	*Ceonothus*; various shrubs	N	Forbes 1954, DFS	N
Metalectra richardsi	Lichens	MV		MV is type locality
Morrisonia mucens	*Quercus*	None	Forbes 1954	
Psaphida thaxteriana	*Quercus*	None	Forbes 1954	
Renia nemoralis	Leaf litter	N, MV	Forbes 1954	MV is approximate northern limit
Xylotype capax	*Vaccinium*	N, MV	Forbes 1954	
Zale sp.†	*Quercus ilicifolia*	NA	DFS	
Zale curema	*Pinus*	N	Forbes 1954	
Zale obliqua	*Pinus*	N	Forbes 1954	
Zanclognatha theralis	Unknown	N, MV	Forbes 1954	
Saturniidae				
Anisota stigma	*Quercus ilicifolia*	MV		1993 Nantucket record (K. Coombs-Beattie pers. comm.)
Eacles imperialis†	*Pinus rigida*	MV		Extirpated from mainland New England
Hemileuca maia†	*Quercus ilicifolia, Q. prinoides*	N, MV		
Grasslands				
Arctiidae				
Cycnia inopinatus†	*Asclepias* spp.; prefers *A. tuberosa*	None		No other known New England occurrences
Grammia phyllira	Unknown; presumably grass	N		Not verified from MV since Jones and Kimball 1943

Putative Primary Habitat Associations	Host Plant(s)	Historical Occurrences	Sources	Comments
Grasslands *(continued)*				
Geometridae				
Semiothisa eremiata†	*Tephrosia virginiana*	N, MV	Schweitzer 1983	Occurs in frost bottoms
Noctuidae				
Abagrotis crumbi benjaminit	Unknown	N, MV		Nominate form occurs in Colorado; on MV occurs in frost bottoms
Apamea burgessi	Probably grasses	N, MV		Occurs in frost bottoms
Cucullia speyeri	*Erigeron canadensis; Aster* spp.	N, MV	Forbes 1954, DFS	
Euxoa violaris†	Unknown	N, MV	Forbes 1954	Occurs in frost bottoms; 1993 Nantucket record (K. Coombs-Beattie pers. comm.)
Faronta rubripennis	Unknown	N, MV	Forbes 1954	SH
Hydraecia immanis	*Humulus lupulus*	MV	Forbes 1954	SH
Lepipolys perscripta	*Linaria canadensis*	N, MV		
Nola sorghiella	*Sorghastrum nutans*	MV	Schweitzer 1983	Stray
Rhodoecia aurantiago	Seedpods of *Aureolaria*	MV		Not verified on MV since Jones and Kimball 1943; possibly a savanna species
Schinia bifascia	Unknown	MV	Forbes 1954	Near northern limit; not recorded from MV since Jones and Kimball (1943)
Heathlands				
Geometridae				
Cingilia catenaria†	*Gaylussacia* and other shrubs	N, MV		Has declined from mainland New England
Eumacaria latiferrugata	*Prunus*	N, MV	Forbes 1948	Occurs in frost bottoms
Glena cognataria	*Vaccinium*	N, MV	Forbes 1948	Occurs in frost bottoms
Noctuidae				
Acronicta lanceolaria	Various shrubs; *Prunus*	MV	Forbes 1954, DFS	Not recorded on MV since Jones and Kimball 1943
Psectraglaea carnosa	*Gaylussacia* suspected	N, MV		Occurs in frost bottoms

Putative Primary Habitat Associations	Host Plant(s)	Historical Occurrences	Sources	Comments
Heathlands *(continued)*				
Sphingidae				
Hemaris gracilis	*Gaylussacia; Kalmia?*	None	Forbes 1948	Rare on mainland New England
Sphinx drupiferarum	*Prunus maritima*	N, MV	Forbes 1954, Schweitzer 1983	Has declined on mainland New England
Wetland/Shoreline				
Geometridae				
Metarranthis pilosaria†	Various Ericaceae	None	Forbes 1948, DFS	
Noctuidae				
Apamea mixta	Probably grasses	None		
Bagisara rectifasciat†	*Hibiscus palustris*	N, MV	Forbes 1954	
Hemipachnobia monochromatea	*Drosera; Kalmia*	MV	Forbes 1954	
Papaipema appassionata†	*Sarracenia*	N	Forbes 1954	Not likely on MV
Papaipema harrisii†	*Heracleum*	MV	Forbes 1954	
Papaipema stenocelis†	*Woodwardia*	N, MV	Forbes 1954	N and MV were first Massachusetts records
Papaipema sulphurata†	*Decodon verticillatus*	N, MV	Bird 1926, Forbes 1954	
Schinia spinosae	Unknown	N, MV		
Spartiniphaga includens	*Spartina* spp.	MV		
Spartiniphaga inops†	*Spartina pectinata*	MV	Forbes 1954	
Notodontidae				
Datana major	*Lyonia, Leucothoe*	MV	Forbes 1948	
Others				
Noctuidae				
Hypocoena defecta	Unknown	MV	Forbes 1954	SH

Invertebrate Response to Insecticide Use: What Happens after the Spraying Stops?

Peter D. Vickery, Jeffrey V. Wells, and E. Richard Hoebeke

Abstract

Spraying of insecticides has obvious short-term effects on insect abundance, but it can also have long-term effects on insect diversity. The Kennebunk Plains, a sandplain grassland in southern Maine, used to be managed for commercial lowbush blueberry (*Vaccinium angustifolium*) production and thus was sprayed annually for more than 20 years (approximately 1950-1987) with the broad-spectrum insecticide Guthion[R]. During the latter half of this period, in 1984, we collected insects using a standard sweep-net method. In 1987 the site was purchased as habitat for endangered plants and animals, and insecticide use ceased. We collected insects again in 1991, four years after insecticide spraying had ceased. Insect diversity and biomass increased significantly in the post-spray period. This suggests that some forms of insects can disperse over wide areas and find and successfully colonize habitat islands fairly rapidly, even when such islands are relatively isolated.

Introduction

Sandplain grasslands in Maine and New England are isolated patches of grassland habitats that support a unique flora and a distinctive invertebrate fauna (Goldstein 1997, Wells et al. 1997). The metapopulation dynamics of invertebrates in these types of grasslands and in many other habitats in the northeastern United States are poorly known and have not been adequately studied (Goldstein 1997). For example, the regal fritillary (*Speyeria idalia*), an obligate grassland specialist once common on grassland habitats from southern Maine to Connecticut and Long Island, NY, has become extirpated in coastal New England since the mid-1970s (Wagner et al. 1997). This extinction underscores the fact that there are essentially no studies concerning the population dynamics of the invertebrates that occupy these unique grassland habitats (Goldstein 1997).

Sandplain grassland sites have become increasingly isolated from each other. In order to develop more meaningful conservation plans that incorporate long-term invertebrate survivorship, then, it is critical to develop some measure of the potential for different groups of invertebrates to recolonize and disperse from these sites. Such information would be essential, for example, in deciding which spatial configuration of reserves would be most likely to preserve the greatest number of invertebrate species.

The Kennebunk Plains, one of two relatively remote sandplain grasslands in southern Maine, was managed for commercial blueberry production from approximately 1950 until 1987. As a result of this agricultural practice, a variety of anthropogenic disturbances took place here for more than two decades, including biennial mowing and burning, extensive use of insecticides (primarily Guthion [azinphos-methyl]), and beginning in 1984, use of herbicides (see Vickery et al. 1992a, b, c). In 1987 the site was purchased by The Nature Conservancy and the State of Maine as habitat for rare plants and animals, and insecticides and herbicides have not been used on the site since then.

The changing management practices associated with different land managers provided us with an unexpected opportunity to learn about rates of invertebrate recolonization at an isolated sandplain grassland site. We had previously collected invertebrates at the Kennebunk Plains in 1984, during the Guthion spray period. We collected invertebrates again in 1991, four years after the last spray event. We sought to learn if any changes in the invertebrate fauna had taken place in the four years since insecticide use. We predicted that invertebrate diversity and biomass would be higher in 1991 than they had been in 1984, and we were especially interested to learn which invertebrates would be found on the site during the post-spray period in 1991.

We made one major assumption in this study: that new invertebrates found in 1991 had recolonized the site rather than established new colonies for the first time. Because of a lack of pretreatment data (the site had been sprayed for more than 20 years), it was impossible to document the invertebrate fauna prior to insecticide use. Thus, the composition and biomass of the "original" invertebrate fauna at the Kennebunk Plains remain unknown. Because the families identified in this study are generally quite common, however, we think our assumption was valid.

Study Site

The Kennebunk Plains is located in Kennebunk, York Co., ME (43°23'N, 70°37'W) and is a 240-hectare (ha) sandplain grassland situated on a broad glacial-marine delta with thick sand deposits. The site supports a xeric native flora dominated by native graminoids (poverty grass [*Danthonia spicata*], little bluestem [*Schizachyrium scoparium*], and Pennsylvania sedge [*Carex pensylvanica*]); shrubs (lowbush blueberry and chokeberry [*Photinia* (formerly *Aronia*) spp.]); and forbs (northern blazing star [*Liatris scariosa* var. *novae-angliae*] and whorled loosestrife [*Lysimachia quadrifolia*]) (see Vickery et al. 1992a for further details).

In 1984, during the first phase of this study, the Kennebunk Plains was being managed for commercial blueberry production. Blueberry production entailed a biennial rotation; approximately 50 percent of the site was mowed and burned each year, and the remaining area was harvested. Guthion, a broad-spectrum insecticide that persists for approximately two weeks before breaking down into constituent components, was applied once annually, in mid-July, to the portion scheduled for harvest. In 1984 this practice had been taking place for more than 20 years.

In 1991, during the second phase of the study, there was no active management for berry production, although there had been prescribed late-fall burns on about half of the site.

Methods

In both 1984 and 1991 we collected invertebrates using a standard sweep-netting technique with a butterfly net (Southwood 1966). The net was flattened on one side to a length of approximately 40 centimeters (cm) so we could capture a higher proportion of nonvolant insects. In each sample collection we used rapid sweeps of approximately 0.5 meters (m) that passed through low vegetation in the same direction, and we swept approximately 0.5 m² for each sample. Because only one individual was responsible for collecting more than 80 percent of the samples, we think there was little bias in our sampling protocol. We recognized that sweep-netting would not capture the full array of invertebrates at the site and that it likely would sample a higher proportion of vagile, volant insects but miss or under-represent terrestrial invertebrates such as spiders (Arachnida) and ants (Formicidae; Southwood 1966).

In 1984 we sampled two unsprayed control plots and four treatment plots. The treatment plots had been aerially sprayed with Guthion at a rate of approximately 280 grams (g) active ingredient/ha on 14 July. The control plots had been sprayed in 1983. We sampled both control and treatment plots in three different periods: pre-spray (between 6 and 12 July), immediately post-spray (between 15 and 20 July), and approximately 5 weeks post-spray (between 20 and 25 August). At least four sweep-net samples were taken from each plot during each period. During the August period we were able to sample only three of the treatment plots.

In 1991, four years post-spray, we collected at least four sweep-net samples from the same plots between 11 July and 6 August. All collections were made under similar sunny conditions; winds were less than 10 kilometers (km) per hour (hr), and temperatures were between 21 and 29°C. Samples were separated from detritus and then placed in a drying oven, set at 60°C for at least 48 hr. Dry weight was measured to 10^{-4} g. Each specimen was identified to family. In the few cases for which family identity could not be established, these samples were known to differ from any previously identified families.

Statistical Analyses

For the 1984 data we used One-way Analysis of Variance (ANOVA) to test for possible differences in invertebrate diversity on control and treatment plots during three separate time periods (pre-spray, immediately post-spray, and five weeks post-spray; Sokal and Rohlf 1981). To determine if possible differences in treatment effects might be an artifact of inherent plot differences prior to Guthion application, we used One-way ANOVA to test if control and treatment plots differed in the pre-spray period. Similarly, we compared control plots during the pre-spray and immediately post-spray periods to see if they had changed. We also compared August control plots (five weeks post-spray) with pre-spray and immediately post-spray control plots to see if invertebrate diversity changed during the latter part of the summer. Similarly, we compared August treatment plots with immediately post-spray treatment plots to determine if invertebrate diversity had changed five weeks post-spray. Finally, to learn if invertebrate diversity had stabilized five weeks after Guthion application, we compared August control plots to August treatment plots.

We used One-way ANOVA to determine if there were differences in invertebrate biomass between all pre-spray plots (control and treatment), the four treatment plots, and the 1991 samples. Because we were not interested in plot differences in the four immediately post-spray plots, we pooled samples for this analysis into each of the following classes: 1984 pre-spray, 1984 immediately post-spray, and 1991 (four years post-spray). We used this same procedure to test for differences in the number of invertebrate families between 1984 pre-spray, 1984 immediately post-spray, and 1991 collections.

We used Principal Components Analysis to help elucidate broader ecological patterns underlying these data (Wilkinson 1990a, b). We plotted 99 percent confidence interval bivariate ellipses of the second and third factors from the Principal Components Analysis. The ellipse is centered on the sample means of the second and third factor loadings (see Wilkinson 1990b for further details). We plotted 1984 pre-spray, 1984 immediately post-spray, and 1991 (four years post-spray) ellipses separately to determine possible relationships among these sampling periods. Because we wanted to see if there were patterns that might emerge from the suite of invertebrates collected during the three different periods, we limited this analysis to samples that actually included invertebrates. Sample sizes for 1984 (both pre-spray and immediately post-spray) and 1991 were equal (n = 23 for each period).

Results

In 1984 there were no differences in invertebrate diversity between control and treatment plots prior to insecticide spray (F = 2.454, df = 1, p = 0.119), and diversity on control plots did not differ in the periods pre- and post-spray (F = 2.14, df = 1, p = 0.237; Fig. 1). Invertebrate diversity declined on treatment plots after spraying (F = 120.764, df = 1, p = 0.000; Fig. 1). Not surprisingly, invertebrate biomass also declined on post-spray treatment plots in 1984 (F = 8.183, df = 1, p = 0.006; Fig. 2).

Interestingly, invertebrate diversity on the treatment and control plots in 1984 had converged by late August, five weeks post-spray. By this time, diversity had declined on control plots compared to pre-spray and immediately post-spray control plots (F = 18.256, df = 1, p = 0.000; Fig. 1). Conversely, invertebrate diversity on treatment plots had increased compared to immediately post-spray treatment in mid-July

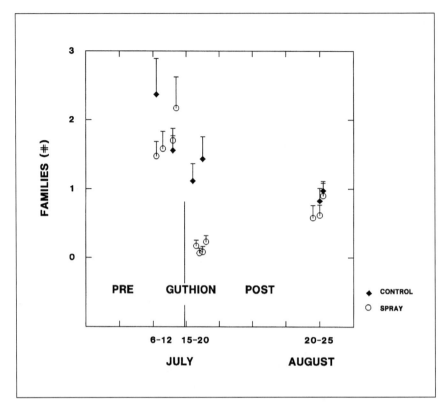

Figure 1. Change in invertebrate diversity (mean ± 1 SE) as a result of Guthion spraying at the Kennebunk Plains, ME, 14 July 1984. The numbers of families declined in treatment plots (circles) immediately after insecticide spraying. By late August, however, the numbers of families in treatment plots had partially recovered and were similar to those in control plots (diamonds).

($F = 12.623$, df = 1, $p = 0.000$; Fig. 1). Invertebrate diversity on control and treatment plots did not differ in late August ($F = 1.168$, df = 1, $p = 0.281$; Fig. 1).

When we compared invertebrate diversity and biomass between 1984 (pre-spray) and 1991 (four years post-spray), we found that both diversity ($F = 15.367$, df = 1, $p = 0.000$) and biomass ($F = 15.619$, $p = 0.000$) had increased (Figs. 2 and 3).

We collected a total of 37 families in 1984 and 1991 (23 families in each time period; Table 1). However, in 1984 only 10 families were found in more than 5 percent of the samples, whereas in 1991, 16 families were found in more than 5 percent of the samples (Table 1). Short-horned grasshoppers (Acrididae), dictyophand planthoppers

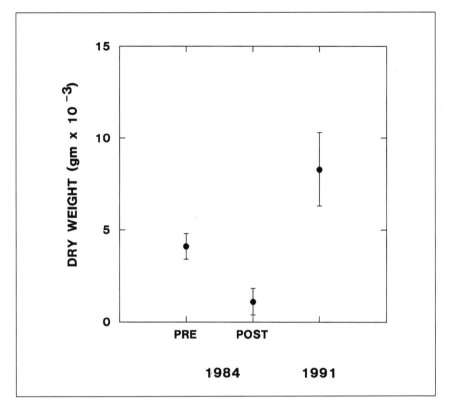

Figure 2. Invertebrate biomass (mean ± 1 SE) declined after Guthion application at the Kennebunk Plains, ME, 14 July 1984. Four years after insecticide use at this site (1991), invertebrate biomass had increased over pre- and post-spray periods (1984).

(Dictyopharidae), aphids (Aphididae), unidentified Lepidoptera, ants (Formicidae), and green metallic bees (Halictidae) were notably more common in 1984 than in 1991. Crickets (Gryllidae), seed bugs (Lygaeidae), planthoppers (Acanaloniidae), leaf beetles (Chrysomelidae), fruit flies (Chloropidae), and three families of Hymenoptera were more common in 1991 than in 1984 (Table 1).

Principal Components Analysis revealed no clear taxonomic patterns (Table 2). The first component was composed largely of families in the orders of Homoptera and Diptera, but the loadings were unclear; both components had positive and negative loadings. The second component was more meaningful and was negatively associated with Hymenoptera and positively associated with Diptera. The third component appeared to reflect diversity; the six families represented five different orders, but none of the families had high values. The fourth

Table 1. Percentage of sweep-net samples that contained a given family taken at the Kennebunk Plains, ME, in 1984 and 1991 (n = 23 for each time period).

Family	1984	1991
Arachnida		
unidentified spider*	0	8.7
Orthoptera		
Acrididae - short-horned grasshoppers	47.8	21.7
Tettigoniidae - long-horned grasshoppers	4.3	0
Gryllidae - crickets	4.3	26.1
Hemiptera		
Miridae - leaf bugs	4.3	0
Nabidae - damsel bugs	4.3	0
Tingidae - lace bugs	0	8.7
Lygaeidae - seed bugs	13.0	26.1
Rhopalidae - scentless plant bugs	4.3	4.3
Pentatomidae - stink bugs	0	4.3
Homoptera		
Cercopidae - spittlebugs	4.3	0
Cicadellidae - leafhoppers	56.5	65.2
Acanaloniidae - planthoppers	0	17.4
Delphacidae - delphacid planthoppers	4.3	0
Dictyopharidae - dictyopharid planthoppers	21.7	4.3
Aphidae - aphids	26.1	0
unidentified Homoptera*	0	4.3
Neuroptera		
Myrmeleontidae - antlions	0	4.3
Coleoptera		
Mordellidae - tumbling flower beetles	8.7	0
Chrysomelidae - leaf beetles	17.4	73.9
Lepidoptera		
Tortricidae - miller moths	0	4.3
unidentified Lepidoptera*	21.7	13.0
Diptera		
Ceratopogonidae - biting midges	4.3	0
Therevidae - stiletto flies	0	4.3
Dolichopodidae - long-legged flies	4.3	0
Syrphidae - syrphid flies	4.3	0
Pipunculidae - big-headed flies	4.3	0
Agromyzidae - leaf miner flies	4.3	0
Lauxaniidae - lauxaniid flies	0	8.7
Chloropidae - fruit flies	0	60.9
Muscidae - muscid flies	0	8.7
unidentified Diptera*	8.7	0

Table 1. continued

Family	1984	1991
Hymenoptera		
Braconidae - parasitic wasps	0	8.7
Ichneumonidae - parasitic wasps	0	8.7
Halictidae - green metallic bees	13.0	0
Formicidae - ants	65.2	8.7
unidentified Hymenoptera*	0	13.0

*Indicates an unidentified family that is known to be different from previously identified families.

Table 2. Principal components loadings for the three highest positive and negative loadings for four axes for invertebrate taxa collected at the Kennebunk Plains, ME, in 1984 and 1991.

Loading 1		Loading 2	
Cercopidae	-0.935	Ichneumonidae	-0.853
Pipunculidae	-0.935	Halictidae	-0.842
Dictyopharidae	-0.902	Rhopalidae	-0.718
Formicidae	0.137	Lauxaniidae	0.144
Acanaloniidae	0.139	Acanaloniidae	0.173
Muscidae	0.133	Muscidae	0.188
Loading 3		**Loading 4**	
Lauxaniidae	-0.526	Nabidae	-0.662
Chloropidae	-0.473	Aphidae	-0.662
Tingidae	-0.336	Formicidae	-0.569
Gryllidae	0.658	Cicadellidae	0.491
Chrysomelidae	0.521	Acanaloniidae	0.480
Tortricidae	0.518	Lygaeidae	0.474

component was composed of Homoptera and Hemiptera, but as with the first component, no clear taxonomic pattern emerged.

Plotting the second and third components, however, revealed an interesting pattern (Fig. 4). Although most samples clustered near the center, 99 percent bivariate ellipses for the 1984 pre-spray period, 1984 post-spray period (comprising both the immediately post-spray and five-weeks-post-spray periods), and 1991 (four years post-spray) were notably different. Ellipses of both 1984 pre-spray and 1984 post-

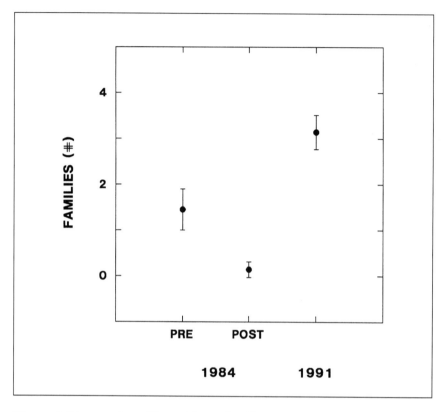

Figure 3. The number of invertebrate families (mean ± 1 SE) declined after Guthion application at the Kennebunk Plains, ME, 14 July 1984. Four years after insecticide use at this site (1991), the number of families had increased over pre- and post-spray periods (1984).

spray invertebrates were similar in size, but the axes for these two periods were almost perpendicular to each other. The large size of the 1991 ellipse revealed the broader taxonomic diversity of these samples. This does not necessarily mean that more families were collected in 1991 than in 1984, but it does indicate that 1991 samples were more likely to include a broader array of families and were less likely to reflect the simplicity underlying the 1984 samples.

Discussion

Despite the fact that this study was constrained in several ways, the results provide important evidence regarding the dispersal and recolonization abilities of some invertebrate taxa. Except for one family, an unidentifiable spider (Arachnida) family, the increased

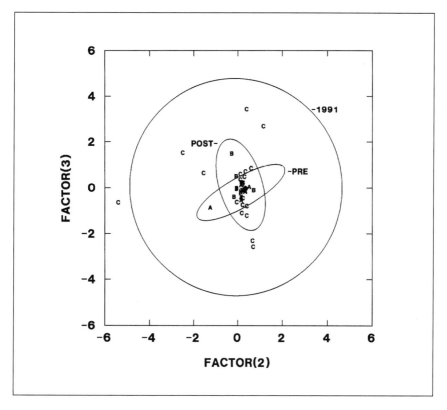

Figure 4. Ninety-nine percent confidence ellipses of the second and third component loadings from a Principal Components Analysis of invertebrate families collected at the Kennebunk Plains, ME, in 1984 and 1991. Ellipses of the 1984 pre-spray (A) and 1984 post-spray (B) periods clustered near the center but in different axes. The large 1991 ellipse (C) indicates a broader diversity in the invertebrate samples taken during this period.

diversity found in the 1991 samples represented vagile families of invertebrates with highly developed dispersal capabilities. Although some northeastern grasslands have persisted for hundreds of years (Winne 1988), many grassland habitats are spatially and temporally ephemeral, and thus the ability to recolonize grassland habitats should be highly adaptive. One would predict that grassland-specialist invertebrates would have excellent dispersal capabilities.

However, the increasing isolation of grassland habitat patches in the northeastern United States in the past 150 years (Litvaitis 1993) is likely to have decreased the probability of recolonization of sites even by invertebrates with superior dispersal capabilities. Without more detailed information on the distribution of individual species across the landscape, it is impossible to know the source of colonizing individuals at the Kennebunk Plains. Developing lists of the species found at individual sites, like the list developed for the Kennebunk Plains (Wells et al. 1997) and the lists of invertebrates that The Nature Conservancy is compiling at other sites, is an important step in understanding which sites may serve as sources for certain species.

For some remnant-dependent species such as the regal fritillary, it is already clear that the spatial configuration, size, and quality of grassland habitat patches have been inadequate to prevent extinction in New England (Panzer et al. 1995, Wagner et al. 1997). It is unclear how many grassland specialists have disappeared from northeastern grasslands, but research on Martha's Vineyard, MA, suggests that the number may be considerable (Goldstein 1997).

Our work suggests that for many common and less specialized invertebrate species, the current network of grassland habitats in southern Maine is adequate for maintenance of populations. Thus, a site where a species has been extirpated can be successfully recolonized from unknown source populations. It is unclear how many sites can disappear before this capacity for "rescue" diminishes or is no longer possible. The Kennebunk Plains is currently managed for its biodiversity and should remain an important regional source for many grassland species. To protect remnant-dependent specialist-species, such sites need to be free from human degradation (Kitahara and Fujii 1994, Panzer et al. 1995).

Clearly, use of prescribed fire as a primary management tool for maintaining grassland habitats can have profound implications for invertebrate population dynamics at sites such as the Kennebunk Plains. The effects will vary widely, depending on the scale, proportion, and timing of grassland burned and the vagility of the species affected. For example, local abundances of larval populations of the microlepidopteran moth *Phalonia hospes* were severely depressed by medium-sized fires (15-30 ha) but recovered to prefire densities within two years; smaller fires (1-3 ha) had little effect on local populations of *Phalonia* larvae (P. Vickery unpubl. data).

The presence of parasitic wasps (Braconidae and Ichneumonidae) in the 1991 samples appears to indicate the development of a more complex invertebrate fauna with higher trophic levels. For successful establishment and persistence, it is likely that these parasitic wasps require a minimum prey density and diversity. Insecticide use was undoubtedly lethal to parasitic wasps, but it is also likely that the reduced prey base caused by insecticide application was an additional constraining factor.

Loss of broad invertebrate diversity on grassland systems can have detrimental effects on some plant species because the interrelationships among adult nectar sources, ovipositing hosts, and the factors that maintain these plants are usually complex (Hammond 1995). Similarly, loss of rare species and breakdown of different trophic levels may lead to wide oscillations in the populations of common species (Ricklefs 1983).

Conclusion

We recognize that this study would have benefited from examining diversity at the species level as well. Clearly, some species within a family have different habitat requirements and are likely to have different dispersal potentials. However, because this study examined recolonization patterns at a higher taxonomic level, it provides insight into the broader patterns of invertebrate dispersal and colonization.

We hope these results will stimulate further research on invertebrate metapopulation dynamics of grassland habitats. Our research highlights the need for further studies in this area, particularly for species in families that are considered to have poor dispersal abilities. A better understanding of these dynamics will be critical to the success of conservation efforts (Hammond 1995).

Acknowledgments

Many people facilitated this research, notably A. Childs, B. Vickery, G. Vickery, and S. Vickery. Warm overnight hospitality and friendship were extended by L. Eastman. Part of this research was supported by the Maine Chapter and Eastern Regional Office of The Nature Conservancy, Maine Audubon Society, Pesticides Control Board of the Maine Department of Agriculture, and Massachusetts Audubon Society. The manuscript benefited from careful review by A. Childs, D. F. Mairs, R. G. Dearborn, and N. Sferra.

Literature Cited

Goldstein, P. Z. 1997. Arthropod assemblages and the management of sandplain grasslands on Martha's Vineyard, Massachusetts. Pp. 217-236 *in* Grasslands of northeastern North America: ecology and conservation of native and agricultural landscapes (P. D. Vickery and P. W. Dunwiddie, eds.). Massachusetts Audubon Soc., Lincoln.

Hammond, P. C. 1995. Conservation of biodiversity in native prairie communities in the United States. J. Kansas Ent. Soc. 68: 1-6.

Kitahara, M., and K. Fujii. 1994. Biodiversity and community structure of temperate butterfly species within a gradient of human disturbance: an analysis based on the concept of generalist vs. specialist strategies. Res. Pop. Ecol. 36: 187-199.

Litvaitis, J. A. 1993. Response of early successional vertebrates to historic changes in land use. Conserv. Biol. 7: 866-873.

Panzer, R., D. Stillwaugh, R. Gnaedinger, and G. Derkovitz. 1995. Prevalence of remnant dependence among the prairie- and savanna-inhabiting insects of the Chicago region. Nat. Areas J. 15: 101-116.

Ricklefs, R. E. 1983. The economy of nature. Chiron Press, New York.

Sokal, R. R., and F. J. Rohlf. 1981. Biometry. 2d ed. W. H. Freeman, New York.

Southwood, T. R. E. 1966. Ecological methods with particular reference to the study of insect populations. Methuen and Co., London.

Vickery, P. D., M. L. Hunter, Jr., and J. V. Wells. 1992a. Use of a new reproductive index to evaluate relationship between habitat quality and breeding success. Auk 109: 697-705.

Vickery, P. D., M. L. Hunter, Jr., and J. V. Wells. 1992b. Is density an indicator of breeding success? Auk 109: 706-710.

Vickery, P. D., M. L. Hunter, Jr., and J. V. Wells. 1992c. Evidence of incidental nest predation and its effects on nests of threatened grassland birds. Oikos 63: 281-288.

Wagner, D. L., M. S. Wallace, J. Boettner, and J. S. Elkinton. 1997. Status update and life history studies on the regal fritillary (Lepidoptera: Nymphalidae). Pp. 261-276 *in* Grasslands of northeastern North America: ecology and conservation of native and agricultural landscapes (P. D. Vickery and P. W. Dunwiddie, eds.). Massachusetts Audubon Soc., Lincoln.

Wells, J. V., R. G. Dearborn, D. F. Mairs, E. R. Hoebeke, N. Sferra, E. C. Roux, P. D. Vickery, and M. A. Roberts. 1997. A preliminary list of the insects of the Kennebunk Plains, Maine. Pp. 251-260 *in* Grasslands of northeastern North America: ecology and conservation of native and agricultural landscapes (P. D. Vickery and P. W. Dunwiddie, eds.). Massachusetts Audubon Soc., Lincoln.

Wilkinson, L. 1990a. Systat: the system for statistics. Systat, Inc., Evanston, IL.

Wilkinson, L. 1990b. Sygraph: the system for graphics. Systat, Inc., Evanston, IL.

Winne, J. C. 1988. History of vegetation and fire on the Pineo Ridge blueberry barrens in Washington County, Maine. Master's thesis, Univ. of Maine, Orono.

Peter D. Vickery: Center for Biological Conservation, Massachusetts Audubon Society, Lincoln, MA 01773, and Department of Forestry and Wildlife Conservation, University of Massachusetts, Amherst, MA 01003.

Jeffrey V. Wells: New York Cooperative Fish and Wildlife Research Unit, Fernow Hall, Cornell University, Ithaca, NY 14853 (current address: National Audubon Society, Cornell Laboratory of Ornithology, 159 Sapsucker Woods Road, Ithaca, NY 14850).

E. Richard Hoebeke: Department of Entomology, Cornell University, Ithaca, NY 14853.

A Preliminary List of the Insects of the Kennebunk Plains, Maine

Jeffrey V. Wells, Richard G. Dearborn, Donald F. Mairs, E. Richard Hoebeke, Nancy Sferra, Eric C. Roux, Peter D. Vickery, and Michael A. Roberts

This is a preliminary list of the insect species that have been recorded at the Kennebunk Plains in southern Maine. Most records are from work done from 1992 to 1995, but butterfly sight records and a few specimen records date from as early as 1984. We did most of our collecting either with a sweep net or by capturing insects by hand. However, we also set light traps at night several times in 1994 and specialized traps for necrophagous species over a three-day period in 1995. Since collecting opportunities were largely incidental to other activities, we have no detailed record of our collecting effort, and therefore this list should not be taken as an index of total species diversity at the site. Some standardized collecting has been done at the Kennebunk Plains (Vickery et al. 1997), but very few specimens from those collections have been identified to the species level.

The Kennebunk Plains is a 240-hectare (ha) sandplain grassland dominated by a mixture of lowbush blueberry (*Vaccinium angustifolium*) and various grasses (little bluestem [*Schizachyrium scoparium*], poverty grass [*Danthonia spicata*], and sheep fescue [*Festuca ovina*]). This habitat is rare in Maine and supports several unusual vertebrate and plant species, including Grasshopper Sparrow (*Ammodramus savannarum*), black racer (*Coluber constrictor*), and northern blazing star (*Liatris scariosa* var. *novae-angliae*). Essentially no information has been available about the invertebrate species found in this habitat. Therefore, we began a project to collect and identify the insect species that occur at the Kennebunk Plains. Most specimens are housed in the Maine Forest Service Entomological Collection in Augusta, ME. Some Odonata and Lepidoptera specimens are in the reference collection of the Maine Chapter of The Nature Conservancy in Brunswick, ME, and some Coleoptera specimens are in E. C. Roux's personal collection.

Although we have not completed an exhaustive search of the literature and major collections, several species that we identified apparently had not been recorded previously in Maine. Several coleopterans, for example, were new to the Maine Forest Service entomological collection (Dearborn and Donahue 1993). Many of these species are typically found in dry grassland habitats in southern New England and therefore might have been reasonably suspected of occurring at the Kennebunk Plains. We have noted species for which we are currently unaware of previous state records. Although 10 species are thought to be first records in Maine, it is likely that some of these species occur at several southern Maine localities.

Key to Letter Codes
S = identified by sight only
C = identified in hand after collected
V = voucher specimen preserved

Order Odonata
Family Coenagrionidae
    Enallagma spp. (bluets) — S
    Nehalennia irene (sedge sprite) — S
    Ischnura verticallis (common forktail) — C, V
Family Aeshnidae
    Anax junius (common green darner) — S
Family Gomphidae
    Gomphus exilis — C, V
Family Corduliidae
    Epitheca cynosura (common baskettail) — C, V
Family Macromiidae
    Didymops transversa (stream cruiser) — C, V
Family Libellulidae
    Celithemis elisa (Elisa skimmer) — C
    Leucorrhinia hudsonicus (whiteface) — C, V
    Libellula julia — C
    Libellula lydia (whitetail) — S
    Libellula pulchella (tenspot) — C
    Sympetrum costiferum (saffron-bordered meadowfly) — C, V

Insects of Kennebunk Plains, Maine

Order Orthoptera
Family Tettigoniidae (long-horned grasshoppers)
    Scudderia pistillata — C, V
    Scudderia curvicauda — C, V
    Conocephalus fasciatus — C, V
Family Oecanthidae
    Oecanthus quadripunctatus — C, V
Family Gryllidae
    Gryllus sp. — C, V
Family Tetrigidae
    Nomotettix c. cristatus — C, V
Family Acrididae (short-horned grasshoppers)
    Arphia sulphurea — C, V
    Arphia xanthoptera — C, V
    Dissosteira carolina (Carolina locust) — C
    Melanoplus s. sanguinipes — C, V
    Melanoplus bivittatus — C, V
    Melanoplus f. femurrubrum — C, V
    Melanoplus fasciatus — C, V
    Melanoplus keeleri luridus — C, V
    Melanoplus confusus — C, V
    Chorthippus curtipennis — C, V
    Orphulella speciosa — C, V
    Pseudopomala brachyptera — C, V
Family Mantidae
    Mantis religiosa (European mantid) — C
Family Blattellidae
    Parcoblatta virginica (northern wood roach) — C, V

Order Hemiptera
Family Lygaeidae
    Cymus sp. — C, V
    Phlegyas abbreviatus — C, V
Family Tingidae
    Corythucha cydoniae (hawthorn lacebug) — C, V

Order Homoptera
Family Delphacidae
    Liburniella ornata — C, V

Order Neuroptera
Family Corydalidae
    Chauliodes pectinicornis — C
Family Myrmeleontidae (antlions)
    Hesperoleon abdominalis — C, V (probable first Maine record)

Order Coleoptera
Family Cicindelidae
    Cicindela punctulata — C, V
Family Carabidae
    Calosoma calidum — C, V
    Calosoma sycophanta — C, V
    Harpalus sp. — C, V
    Lebia viridis — C, V
Family Hydrophilidae
    Hydrochara sp. — C, V
Family Silphidae
    Nicrophorus orbicollis — C, V
    Nicrophorus tomentosus — C, V
    Nicrophorus defodiens — C, V
    Nicrophorus marginatus — C, V
    Necrodes surinamensis — C, V
    Necrophila americana — C, V
    Oiceoptoma noveboracense — C, V
Family Staphylinidae
    Staphylinus vulpinus — C, V
    Creophilus maxillosus — C, V
    Ontholestes cingulatus — C, V
Family Histeridae
    Euspilotus assimilis — C, V
Family Cleridae
    Enoclerus muttkowskii — C, V (probable first Maine record)
    Phyllobaenus pallipennis — C, V
Family Elateridae
    Athous rufifrons — C, V
    Ctenicera sjaelandica — C, V (probable first Maine record)
Family Lampyridae
    Lucidota nigricans — C, V
Family Buprestidae
    Dicerca sp. — C, V
Family Scarabaeidae
    Copris fricator — C, V
    Onthophagus hecate — C, V
    Onthophagus nuchicornis — C, V
    Geotrupes splendidus — C, V
    Cotalpa lanigera (goldsmith beetle) — C, V
    Aphonus tridentatus — C, V
    Phyllophaga gracilis — C, V

Insects of Kennebunk Plains, Maine

Order Coleoptera *(continued)*
Family Scarabaeidae *(continued)*

Phyllophaga fusca	C, V	(probable first Maine record)
Phyllophaga drakii	C, V	(probable first Maine record)
Phyllophaga forsteri	C, V	(probable first Maine record)
Dichelonyx sp.	C, V	
Macrodactylus subspinosus (rose chafer)	S	
Polyphylla variolosa	C, V	

Family Cerambycidae

Tetraopes tetrophthalmus (red milkweed beetle)	C, V	

Family Chrysomelidae

Pachybrachys sp.	C, V	
Cryptocephalus gibbicollis	C, V	(probable first Maine record)
Chrysochus auratus (dogbane beetle)	S	
Altica sylvia (blueberry flea beetle)	C, V	

Order Lepidoptera
Family Coleophoridae

Coleophora sp.	C, V	

Family Blastobasidae

Blastobasis sp.	C, V	
Hypatopa sp.	C, V	

Family Gelechiidae

Isophrictis sp.	C, V	

Family Pyralidae

Scoparia sp.	C, V	
Acrobasis betulella (birch tubemaker)	S	
Acrobasis comptoniella (sweetfern tubemaker)	S	
Hypsopygia costalis (clover hayworm moth)	C, V	
Agriphila vulgivagella (vagabond crambus)	C, V	
Crambus leachellus	C, V	
Herpetogramma thestealis	C, V	

Family Sesiidae (= Aegeriidae)

Carmenta anthracipennis	C, V	(probable first Maine record)

Family Tortricidae

Phalonia hospes	C, V	
Phytcholoma peritana	C, V	
Sparganothis sulfureana (sparganothis fruitworm moth)	C	
Phaneta sp., possibly *ochrocephala*	C, V	
Olethreutes cespitana	C, V	
Eucosma similiana	C, V	

Order Lepidoptera *(continued)*
Family Papilionidae

Papilio canadensis (tiger swallowtail)	S
Papilio polyxenes (black swallowtail)	S

Family Pieridae

Colias philodice (clouded sulphur)	S
Colias eurytheme (alfalfa butterfly)	C
Colias interior (pink-edged sulfur)	C

Family Lycaenidae

Lycaena phlaeas americana (American copper)	S
Satyrium titus (coral hairstreak)	S
Incisalia augustinus (brown elfin)	C, V
Celastrina argiolus (spring azure)	S

Family Nymphalidae

Speyeria cybele (great spangled fritillary)	S
Speyeria aphrodite or *atlantis* ?	C, V
Boloria bellona (meadow fritillary)	S
Nymphalis antiopa (mourning cloak)	S
Vanessa atalanta rubria (red admiral)	S
Vanessa cardui (painted lady)	S
Junonia coenia (buckeye)	S
Limenitis arthemis (white admiral)	S
Limenitis arthemis astyanax (red-spotted purple)	S
Limenitis archippus (viceroy)	S
Coenonympha tullia inornata (inornate ringlet)	C
Enodia anthedon (northern pearly eye)	S
Megisto cymela (little wood satyr)	S
Cercyonis pegala alope (common wood nymph)	S
Danaus plexippus (monarch)	S

Family Geometridae

Eumacaria latiferrugata (brown-bordered geometer)	C, V
Petrophora subaequaria (northern petrophora)	C, V
Prochoerodes transversata (large maple spanworm moth)	C, V
Pleuroprucha insularia (common tanwave)	C, V

Family Saturniidae

Actias luna (luna moth)	S
Hyalophora cecropia (cecropia moth)	C, V
Dryocampa rubicunda (rosy maple moth)	C, V

Insects of Kennebunk Plains, Maine

Order Lepidoptera *(continued)*
Family Sphingidae
    Hemaris diffinis (snowberry clearwing)   C, V
    Pachysphinx modesta (big poplar sphinx)   C, V
    Smerithus cerisyi (one-eyed sphinx)   C, V
    Sphinx poecilius   C, V
    Poanes excaecatus (blinded sphinx)   C, V
Family Arctiidae
    Crambidia pallida   C, V
Family Lymantriidae
    Lymantria dispar (gypsy moth)   C, V
Family Notodontidae
    Nadata gibbosa (white-dotted prominent)   C, V
Family Noctuidae
    Euxoa mimallonis   C, V
    Protolampra brunneicollis (brown-collared dart)   C, V
    Drasteria occulta (occult drasteria)   C, V
    Lacinipolia meditata (the thinker)   C, V
    Polia purpurissata (purple arches)   C, V
    Noctua pronuba (large yellow underwing)   C, V
    Amphipyra tragopoginis   C, V
    Crymodes devastator (glassy cutworm moth)   C, V
    Calophasia lunula   C, V
    Anagrapha falcifera (celery looper moth)   C, V
    Feltia jaculifera (dingy cutworm moth)   C, V
    Abagrotis placida   C, V
    Schinia lynx   C, V

Order Diptera
Family Therevidae
    Psilocephala haemorrhoidalis   C, V
Family Asilidae
    Proctacanthus philadelphicus   C, V

Order Hymenoptera
Family Argidae
    Arge sp.   C, V
Family Ichneumonidae
    Ichneumon eximus   C, V
Family Pelecinidae
    Pelecinus polyturator   S
Family Mutillidae (velvet ants)
    Dasymutilla vesta   C, V   (probable first Maine record)

Order Hymenoptera *(continued)*	
Family Formicidae	
Polyergus lucidus	C, V (probable first Maine record)
Family Vespidae	
Polistes sp. (paper wasp)	S
Dolichovespula maculata (baldfaced hornet)	S
Family Sphecidae	
Sphex ichneumoneus (great golden digger)	S
Family Halictidae	
Augochlorella striata	C, V
Family Apidae	
Apis mellifera (honeybee)	S

Acknowledgments

We thank the many individuals, including D. Ouellette, M. Russo, D. Schweitzer, A. C. Wells, and D. Wagner, who helped collect and identify specimen materials from the Kennebunk Plains. D. Wagner's comments improved the manuscript notably.

Literature Cited

Dearborn, R. G., and C. P. Donahue. 1993. The forest insect survey of Maine. Order Coleoptera. Tech. Rep. no. 32, Maine Forest Serv., Dept. Conservation, Insect and Disease Div., Augusta.

Vickery, P. D., J. V. Wells, and E. R. Hoebeke. 1997. Invertebrate response to insecticide use: what happens after the spraying stops? Pp. 237–250 *in* Grasslands of northeastern North America: ecology and conservation of native and agricultural landscapes (P. D. Vickery and P. W. Dunwiddie, eds.). Massachusetts Audubon Soc., Lincoln.

Jeffrey V. Wells: New York Cooperative Fish and Wildlife Research Unit, Fernow Hall, Cornell University, Ithaca, NY 14853 (current address: National Audubon Society, Cornell Laboratory of Ornithology, 159 Sapsucker Woods Road, Ithaca, NY 14850).

Richard G. Dearborn: Entomology Laboratory, Maine Forest Service, Department of Conservation, 50 Hospital Street, Augusta, ME 04330.

Donald F. Mairs: Maine Department of Agriculture, Food and Rural Resources, Division of Plant Industry, State House Station 28, Augusta, ME 04333.

Insects of Kennebunk Plains, Maine

E. Richard Hoebeke: *Department of Entomology, Cornell University, Ithaca, NY 14853.*

Nancy Sferra: *The Nature Conservancy, Sanford, ME 04073.*

Eric C. Roux: *Biology Department, Northeastern University, Boston, MA 02135.*

Peter D. Vickery: *Center for Biological Conservation, Massachusetts Audubon Society, Lincoln, MA 01773, and Department of Forestry and Wildlife Conservation, University of Massachusetts, Amherst, MA 01003.*

Michael A. Roberts: *367 Village Road, Steuben, ME 04680.*

Status Update and Life History Studies on the Regal Fritillary (Lepidoptera: Nymphalidae)

David L. Wagner, Matthew S. Wallace, George H. Boettner, and Joseph S. Elkinton

Abstract

Eastern populations of the regal fritillary (*Speyeria idalia*) have been declining since at least the late 1940s. New England populations disappeared from north to south, with the last viable colonies occurring on island sandplain grasslands and heathlands off the coasts of Rhode Island and Massachusetts. Extensive field surveys in 1992 and subsequent efforts have failed to locate any colonies in New England. Females are perhaps the most fecund of all butterflies. In this study, eight captive females laid more than 1,300 eggs per female (range 227-2,494), and egg hatch ranged from 19 to 78 percent. The addition of less than 5 percent raw albumin to the honey-water diet of three adult females in 1993 coincided with an increase in daily egg output and hatch. One cohort of laboratory-bred larvae was almost entirely lost to a nuclear polyhedrosis virus. Young violet (*Viola*) leaves were suitable for the establishment of first instars, but mature foliage was not; the later resulted in 100 percent mortality of first instars.

We discuss reasons for the regal fritillary's decline and make management recommendations. Because females frequently oviposit away from the host plant, dense violet colonies should be especially advantageous for the establishment of the minute first-instar larvae. Given the long life of adults and their propensity for nectar, we think that the availability of late-summer nectar will be essential in efforts to maintain or reestablish this striking insect.

Introduction

With its fiery orange forewings, iridescent blue-black hindwings, and 10-centimeter (cm) wingspan, the regal fritillary was one of New England's most magnificent grassland insects. Although always locally distributed, it formerly occurred in all six New England states (Scudder 1889, Denton 1900, Klots 1951, Opler 1983, 1992). Preferred habitats included marshes and swamp edges, wet meadows, fields, pastures, and native grasslands (Denton 1900, Weed 1926, Clark 1932, Clark and Clark 1951, Klots 1951, Schweitzer 1992, Glassberg 1993; see also Fig. 1). An examination of the numbers of specimens in collections and of accounts in the literature indicates that the species reached its greatest abundance in New England in Massachusetts on the sandplain grasslands of Martha's Vineyard, Nantucket Island, and other offshore islands (Scudder 1889, Jones and Kimball 1943). On the mainland and on Block Island, RI, the species was commonly associated with wet meadows and human-maintained grasslands created by mowing, grazing, and other agricultural practices. As in other *Speyeria*, the larvae are violet specialists. New England's regal fritillary colonies were associated with ovate-leaved violet (*Viola fimbriatula*), lance-leaved violet (*V. lanceolata*), common blue violet (*V. papilionacea*), and birdfoot violet (*V. pedata*) (Schweitzer 1987, Cassie et al. in press, D. Schweitzer pers. comm.).

Adult males begin emerging in mid- to late June, followed one to two weeks later by the first females (Scudder 1889, Clark and Clark 1951, Barton 1993, 1994). Adults are long-lived, with some individuals surviving up to 90 days in the wild (Barton 1993). Although females are believed to pair shortly after emerging from the chrysalis, few eggs are laid until the latter half of August or early September. Eggs hatch after two to three weeks, and the first instars immediately enter diapause, presumably without eating (Edwards 1879, Maynard 1886, Scudder 1889, Weed 1926, Mattoon et al. 1971). Feeding begins in spring, and larvae complete their development by late May or June; the pupal stage lasts an additional two-and-a-half to four weeks (Edwards 1879, Maynard 1886, DLW unpubl. data).

It is not known exactly when the regal fritillary began its decline in New England, but a letter from amateur lepidopterist John Bakeless in the correspondence of Alexander B. Klots (Homer Babbidge Library, Special Collections, Univ. of Connecticut) indicated that viable colonies in Connecticut were already disappearing by the late 1940s. The last known Connecticut colony, located on the Roxbury-Bridgewater town line, disappeared in 1971 (R. Muller pers. comm.). The species' demise evidently proceeded from north to south, with the last mainland sightings occurring in the 1970s and early 1980s. By the mid-1980s only six populations remained, all on offshore islands: Block and

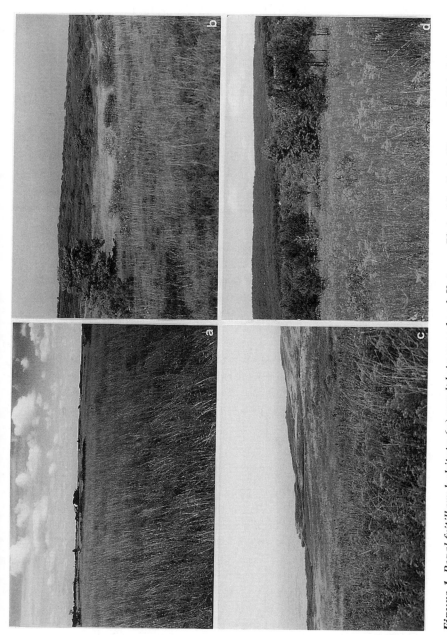

Figure 1. Regal fritillary habitats: (a) sandplain grassland, Katama Plain, Martha's Vineyard, MA; (b) heathland, Nantucket I., MA; (c) coastal pasture, Block Island, RI; and (d) upland grassy meadow, central Pennsylvania.

Conanicut islands in Rhode Island and Martha's Vineyard, Nantucket, Naushon Island, and No Mans Land in Massachusetts (Schweitzer 1987, D. Schweitzer unpubl. data). This pattern of disappearing from north to south was repeated in New York, where the last populations were recorded in the 1980s on Long Island. The last New England regal fritillary was seen by lepidopterist Larry Gall in 1991 on the north end of Block Island. Focused searches of the six aforementioned islands in July and August 1992 yielded no sightings, nor have any been reported since.

In December 1991 a meeting was organized by Scott Melvin, an endangered-species biologist with the Massachusetts Division of Fisheries and Wildlife, to review the status of the regal fritillary in New England and New York. The working group included staff from the U.S. Fish and Wildlife Service, Massachusetts Division of Fisheries and Wildlife, Natural Heritage Programs in Massachusetts, Rhode Island, and Connecticut, The Nature Conservancy, and Massachusetts Audubon Society; university entomologists; and other informed parties. The group recommended a two-step plan of action: first, to establish a captive breeding protocol (if any remaining butterflies were found, genetic stock from indigenous populations could be used to reestablish the species); and second, to study management issues. Only later did participants learn that the regal fritillary was already extirpated from New England and that breeding stock would have to be acquired from outside the region.

Below we discuss results from our efforts in 1992–1993 and 1993–1994 to establish rearing protocols for this butterfly and possible reasons for the species' decline. We discuss briefly two management issues: violet density and nectar availability.

Methods

Females were obtained from colonies in Pennsylvania (1992; $n = 3$), Missouri (1993; $n = 2$), and Iowa (1993; $n = 3$). Each female was individually housed in a 12-liter (l) cardboard ice-cream carton covered with netting and placed on a bench near a south- or east-facing window. Dried violet leaves and strips of crumpled toweling were provided as ovipositional substrates, and containers were misted with water two or three times a week to simulate rainfall. Females were removed and fed once a day with a sugar or honey solution using the feeding stations of Mattoon et al. (1971). Once a female had repeatedly retracted her proboscis or 15 minutes had passed, she was gently rinsed off with water and returned to her container. In 1992 a small portion of raw chicken-egg albumin (less than 5 percent by volume) was added to the diet at the onset of egg-laying; in 1993 this was done 20 and 22 days after oviposition had commenced.

Newly deposited eggs were collected every day, placed in plastic vials, and misted once a week to elevate humidities. Eclosion was monitored daily; newly hatched first instars were transferred to sterilized, moistened wooden blocks, placed in a plastic bag, and held in a refrigerator at 2 to 5° C (Mattoon et al. 1971). Blocks were remoistened every six to eight weeks. Larvae in moldy blocks were transferred to blocks that had been autoclaved.

In the hope of establishing a laboratory colony that would produce multiple generations each year, we prevented first instars from entering diapause during the winter by exposing them to long day lengths and warm temperatures (Mattoon et al. 1971). Diapausing first instars were placed on new violet leaves in glass petri plates with moistened filter paper and held below an incandescent light to raise temperatures to 25 to 30° C for 14 hours each day. The filter paper was kept moist to ensure that the humidity remained high, and a new violet leaf was added daily. Larvae exposed to this regimen usually began feeding within a few hours to two days (a small percentage of stragglers took up to 20 days to begin feeding). Larvae fed on all four of the violets we offered: ovate-leaved violet, common blue violet, birdfoot violet (all of which grow wild in New England), and domestic cultivars of the three-colored violet (*V. tricolor*).

Adult offspring from midwestern stock obtained in 1993 were successfully paired in glass enclosures in January 1994. A second generation of larvae was obtained in February, and this cohort was immediately "forced" through to the adult stage.

Two species of violets associated with regal fritillary populations in New England, birdfoot and ovate-leaved violets, were used in a host and leaf preference study. Wild-collected plants of these two species and other violets were grown in the greenhouse at the University of Connecticut. Larval establishment and survivorship were followed for three size-age classes of leaves. For ovate-leaved violet, size-age classes were assigned as follows: young leaves were 1 to 3 cm long, pale green, and had the bases of the leaves curled over the midrib; intermediate leaves were 2 to 4 cm long, had begun to darken, and were mostly uncurled; and mature leaves were 2 to 6 cm long and were fully darkened and expanded. For birdfoot violet, leaf ages were assigned as follows: young leaves were less than 15 millimeters (mm) long and still had the lobes drawn together; intermediate leaves were 15 to 20 mm long, had begun to darken, and had edges that were partially uncurled and not touching the midrib; and mature leaves were fully darkened with the lobes separated. Rearing was done in glass petri plates to which moistened filter paper had been added. Fresh leaves were provided daily, although a portion of the older leaves was always left as well.

Results

The three regal fritillary females collected 20 August 1992 in Pennsylvania laid an average of 1,447.6 eggs (range 906-1,849); the two females captured in Missouri in late August 1993 laid an average of 1,417.5 eggs (range 341-2,494); and the three females captured in Iowa in late August and early September 1993 laid an average of 1,159.3 eggs (range 227-2,240). The females laid eggs on the sides of the cardboard containers, nylon screening, crumpled paper, and dried violet leaves. They were especially apt to place ova in protected crevices and seams.

Captive females laid eggs over a four- to six-week period, with peak production occurring in the first four weeks of ovipositional activity (Fig. 2A). After the first week in October, egg production was modest. The last eggs were laid on 14 October (1992) and 28 October (1993). All females died within 10 days of their last ovipositional date.

We noted marked fluctuations in daily egg production in both years of the study. In the second year of the study, albumin was added to the diet of one female on day 20 (18 September 1993) and to the diet of two other females on day 22 (20 September 1993). This addition was followed by an increase in fecundity (Fig. 2B). Egg hatch from our eight captive females ranged from 19 to 78 percent (summing across all ovipositional dates for a given female). Percentage egg hatch was similar for our three collections: Iowa 65 percent ($n = 2$), Missouri 62 percent ($n = 2$), and Pennsylvania 65 percent ($n = 3$). Healthy eggs were white or cream colored when laid and darkened to a frosted gray as the caterpillars matured. Eggs that failed to hatch were often yellow and collapsed. The last eggs to be laid were often inviable. As in egg production, we noted considerable fluctuations in egg hatch. In 1993 an increase in hatch coincided with the addition of albumin to the female's honey-water diet, although this observation was largely due to the performance of a single female (Fig. 3).

In 1983, larval development and survivorship were followed for a single cohort of 86 larvae that were started on common blue violet shortly after the larvae emerged from eggs in October. The first five instars were of similar duration, lasting from an average of 3.6 days to 6.6 days. The sixth and final larval instar was much longer, lasting an average of 17.1 days.

After noting that early instars seemed to prefer leaves that were not yet fully expanded, we set up an experiment to examine larval establishment and survivorship on leaves representing three size-age classes

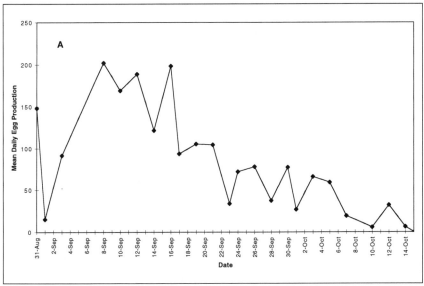

A.

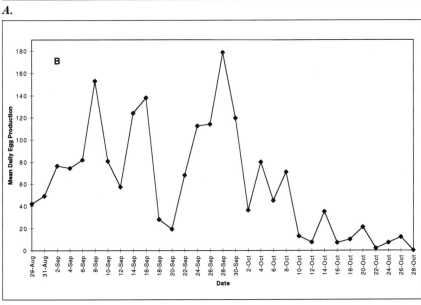

B.

Figure 2. *Regal fritillary oviposition. (A) Mean daily egg production for three Pennsylvania-captured females (1992); albumin was added to their diet from the first day of oviposition. Two females were laying through the first two dates; three females from 3 Sept. until 7 Oct.; two from 7 to 12 Oct.; and only one thereafter. (B) Mean daily egg production for three midwestern-captured females (1993); albumin was added to their diets on 18 and 20 Sept. One female was laying through the first two dates; three females from 2 Sept. until 6 Oct.; two from 6 to 14 Oct.; and only one thereafter.*

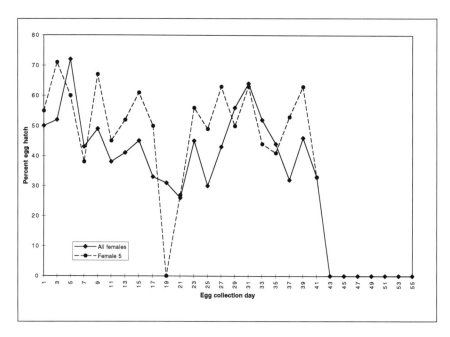

Figure 3. Percentage egg hatch by date for five midwestern-captured regal fritillaries (1993): x axis gives number of days after a given female began depositing eggs. By day 9, one female had stopped laying; by day 19, two had stopped; albumin was added to the diet of the three surviving butterflies on days 20 and 22. By day 39, two females were still laying; by day 41, only a single female was laying (and all of her eggs failed to hatch).

from ovate-leaved and birdfoot violets. Both young and intermediate leaves proved suitable for larval establishment, but no larvae survived more than 13 days on mature leaves of either violet (Table 1). Larvae often remained and continued to feed on one- to three-day-old leaves even after they had wilted.

Discussion

The 2,494 eggs laid by one of our captive regal fritillaries from Missouri were more than twice the previous number known to be laid by any butterfly species (1,200 by an Edith's checkerspot [*Euphydryas editha*]; Labine 1968). It is possible that even greater fecundities might have been realized had our captive females been given access to living flowers, soil exudates, and other natural substrates. Compared with other butterflies, the regal fritillary appears to be a "sweepstakes strategist" that has placed a premium on the number of eggs it can produce. However, not all eggs hatch. For the eight wild-caught

Table 1. *Establishment of first-instar regal fritillary larvae on ovate-leaved violet and birdfoot violet. Stock from Pennsylvania (1992) females. Ten larvae were started on each of the three leaf-classes: young, intermediate, and mature.*

Number Surviving	Day 1	Day 5	Day 9	Day 13
ovate-leaved violet				
young leaves	10	5	4	4
intermediate leaves	10	10	9	9
mature leaves	10	4	0	0
birdfoot violet				
young leaves	10	5	5	5
intermediate leaves	10	6	5	3
mature leaves	10	4	1	0

females in our study, the hatch rate was never more then 78 percent and was as low as 19 percent (mean = 64.3%). We cannot explain why females would invest in the production of such a high percentage of eggs that fail. Nor do we know if the unhatched eggs that we observed were infertile or failed to hatch for some other reason. Although adult nutrition of our captive females or improper incubation conditions may explain our low hatch, similar levels of egg failure have been observed in the Mormon fritillary (*Speyeria mormonia*; C. L. Boggs pers. comm.).

It is likely that most mortality occurs in the first instar, which must survive a six- to seven-month diapause having fed only on a portion of its egg chorion (Edwards 1879, Scudder 1889, Weed 1926, Mattoon et al. 1971). Our lab-reared larvae consumed only enough of the chorion to free themselves from the egg. Females are indiscriminate egg layers, ovipositing on grass, soil, and under pebbles, presumably in the vicinity of violets but only rarely placing eggs on violets (Scudder 1889, Clark 1932, Klots 1951, Scott 1986, Barton 1993, B. Barton pers. comm.); thus, the task of locating new violet leaves in the spring must be difficult for the tiny first instars. Flooding, mold, desiccation, predators, and starvation undoubtedly take high tolls, and the high fecundity of the regal fritillary helps compensate for the losses suffered in the first instar. Even when first-instar larvae were supplied with freshly picked foliage, losses were high (50%) in our lab colony (Table 2).

It was exceedingly difficult in our study to get captive-bred adults to pair in the winter. Adults housed in a flight cage (2 x 2 x 2 meters [m]) and provided with high temperatures and humidities, natural light

Table 2. Larval development of the regal fritillary. A cohort of 86 larvae was forced out of diapause in early October 1993 (see text). An incandescent lamp was used to raise rearing temperatures to approximately 25–27° C.

	N	Mean (d)	SD (d)	Stage Mort[1] (%)	Cumulative Survival (%)
1st instar	86	4.8	3.6	50	50
2d instar	43	5.9	1.6	0	50
3d instar	43	4.1	1.3	5	48
4th instar	41	3.6	1.2	0	48
5th instar	41	6.6	1.7	0	48
6th instar	41	17.1	2.7	37	32
Pupa	26	—	—	31	21
Adult[2]	18	—	—	—	21

[1]Percent succumbing in a given instar.
[2]Surviving at least to pharate adult.

supplemented with artificial lights, violets, honey water, nectar plants (buddleia [*Buddleia*] and heliotrope [*Heliotropium*]), animal dung, and salts showed little interest in coupling. Of the 50-plus adults bred in December (1993) and January (1994), only two pairings were obtained over the four weeks the adults were housed together. Our inability to achieve mating in the winter remains the principal hurdle to establishing a captive breeding program for this butterfly.

Reasons for the Regal Fritillary's Decline

The regal fritillary's decline has been noted rangewide. There is only one viable eastern colony, in central Pennsylvania; the range of the midwestern populations also appears to be contracting along the species' western and northern boundaries (Schildkneckt 1986, Nagel et al. 1991, Opler 1992, Schweitzer 1992, 1993, Glassberg 1993, Swengel 1993). Explanations for the decline include habitat fragmentation and conversion, fire, hurricane impact, pesticide drift, collecting, competition with other *Speyeria*, and the introduction and/or spread of a parasitoid or pathogen (Nagel et al. 1991, Schweitzer 1992, 1993, Swengel 1993). Perhaps all of these factors have contributed to the demise of one or more populations, yet only the first, and possibly the last, are apt to apply rangewide. The regal fritillary was a likely beneficiary of the extensive agriculture practiced across the eastern United States and southern Canada in the early twentieth century. Pastures, for example, are favorable areas for violet growth and have often provided habitat for regal fritillaries, especially when bordering marshy lands or ponds. It is likely that this grassland butterfly peaked in abundance in the Northeast during Colonial times when much of the region had been cleared for crops and pasture (Clark 1914, Bell 1985). Conversely, as New England has reforested, suitable habitat for the butterfly has diminished and become increasingly fragmented.

Nagel et al. (1991) recorded an average daily movement of 0.07 kilometers (km) for 23 recaptured regal fritillaries on the Rowe Sanctuary in Nebraska. Barton (1993, 1994) found that males and females occasionally flew long distances: in 1993 the average distance moved by 22 marked adults identified as dispersers was 3.4 km. One male flew 15.8 km in a 12-hour period. In expansive tallgrass prairies, where the regal fritillary reaches its greatest abundance, such movements would not carry it away from suitable habitat as often as would be expected in the East where grasslands are usually small and isolated by forest. Perhaps the regal fritillary persisted longer on New England's offshore islands because dispersing adults were repeatedly turned back by open ocean.

In the spring of 1994 we lost more than 80 percent of a second-generation cohort of middle- to late-instar larvae to a nuclear polyhedrosis virus (NPV). Prior to this we had seen very low mortalities in second to fifth instars (Table 2). The role that this or other pathogens play in low egg hatch, the deformation of wings, or other aspects of the species' decline remains unknown, but certainly pathogens need to be considered since a natural enemy such as a parasitoid or pathogen could account for the observed pattern of decline. Because pathogens tend to act in a density-dependent fashion, a highly specific agent would be expected to occur at very low incidence or to disappear entirely as host populations dwindled (Anderson and May 1980). If regal fritillaries shared pathogens with other fritillaries or more distantly related Lepidoptera, however, the disease agent could remain abundant and cause continued high mortality, even as regal fritillary populations declined to extinction (see Holt and Lawton 1994).

Reestablishment and Management Considerations: Violets and Nectar

Compared with other butterflies, female regal fritillaries appear to be r-selection strategists with regard to their fecundity and ovipositional behavior. Because females only occasionally oviposit directly onto violets, dense violet growth should promote survivorship of the vulnerable first instars. Tree clearing, mowing, and thatch removal have been employed successfully to favor the proliferation of violets on the coastal prairies where the Oregon silverspot (*Speyeria zerene hippolyta*) has recovered from near extinction (Hammond 1987, 1988, 1989, Hammond and McCorkle 1991). It is important to learn if larval establishment and survivorship are differentially affected by different violet species.

Males and females of all fritillaries are avid nectar feeders (Howe 1975, Dornfeld 1980, Ferris and Brown 1981). In the Mormon fritillary (Boggs and Ross 1993) and zerene fritillary (*Speyeria zerene*; Hammond and McCorkle 1991), adult diet is known to affect both longevity and fecundity. Regal fritillaries feed on nectar throughout their long flight season, from June through September (Barton 1993, 1994). Summer mowings that remove nectar sources such as thistle (*Cirsium*), milkweed and butterfly weed (*Asclepias*), ironweed (*Vernonia*), and others (Scudder 1889, Weed 1926, Nagel et al. 1991, Schweitzer 1992, Barton 1993, 1994) would force adults to move away from emergence sites. Iftner et al. (1992) noted that a colony of regal fritillaries in Ohio disappeared after milkweed was eliminated at the site. In Pennsylvania, an abundance of thistle seems to be especially important for the welfare of the regal fritillary (B. Barton pers. comm.).

Given the descriptive nature of our data, it is difficult to conclude with certainty that the addition of egg albumin to the daily "nectar" solution provided to our regal fritillary females increased fecundity and hatch (Figs. 2B and 3). Even if our results were coincidental, nectar quality and quantity must be essential to the reproductive success of such a long-lived butterfly. Management plans for the regal fritillary should include provisions to ensure the availability of nectar at or near those sites with highest violet densities. Access to nectar, especially in late summer when flowers can be very scarce in New England grasslands, may prove critical in efforts to reestablish or manage this species.

Acknowledgments

G. Leslie and R. Wilson of The Butterfly Place, Westford, MA, generously provided time, effort, space, and resources and eagerly shared their knowledge of butterfly rearing. Many helped with the daily maintenance of our lab colonies: J. Adams, C. Boettner, T. Crowingshield, V. Giles, J. Henry, C. Houle, J. Powell, D. Wakoluk, and G. Witkus. The suggestion to add albumin to the adult diet was made by P. Hammond. B. Barton and D. Schweitzer shared many observations and unpublished data with us. We are indebted to those who helped us obtain gravid females: B. Barton and T. Wilkinson (Pennsylvania), R. Johnson (Missouri), and T. Orwig (Iowa). The figures were prepared by J. Henry. M. Mello located and dug up many of the violets used in the study. Two anonymous reviewers made helpful suggestions on an earlier draft of the manuscript. P. Dunwiddie, R. Enser, P. Goldstein, M. Mello, C. Raithel, T. Simmons, and P. Smith conducted the 1992 field surveys of the six former island colonies. Special thanks to S. Melvin and C. Raithel who sparked us all into action with their wake-up call in 1991. This work was supported by a contract grant (no. 92SMM11 to DLW) and cooperative agreement (no. 14-16-0009-1575 to JSE) from the Massachusetts Division of Fisheries and Wildlife.

Literature Cited

Anderson, R. M., and R. M. May. 1980. Infectious diseases and population cycles of forest insects. Science 210: 658-661.

Barton, B. 1993. Field study of the regal fritillary, *Speyeria idalia*. Report to Department of the Defense and The Nature Conservancy, Harrisburg, PA.

Barton, B. 1994. The status of the regal fritillary. Report to Department of the Defense and The Nature Conservancy, Harrisburg, PA.

Bell, M. 1985. The face of Connecticut: people, geology, and the land. State Geol. Nat. Hist. Surv. Conn. Bull. 110.

Boggs, C. L., and C. L. Ross. 1993. The effect of adult food limitation on life history traits in *Speyeria mormonia* (Lepidoptera: Nymphalidae). Ecology 74: 433-441.

Cassie, B., C. W. Leahy, and R. K. Walton, eds. In press. The butterflies of Massachusetts and other New England states. Massachusetts Audubon Soc., Lincoln.

Clark, A. H. 1932. The butterflies of the District of Columbia and vicinity. U.S. Natl. Mus. Bull. 157.

Clark, A. H., and L. F. Clark. 1951. The butterflies of Virginia. Smithson. Misc. Collect. vol. 116, no. 7.

Clark, G. L. 1914. A history of Connecticut. G. P. Putnam's Sons, New York.

Denton, S. F. 1900. Moths and butterflies of the United States east of the Rocky Mountains. Bradlie Whidden, Boston.

Dornfeld, E. J. 1980. The butterflies of Oregon. Timber Press, Forest Grove, OR.

Edwards, W. H. 1879. Description of the preparatory stages of *Argynnis idalia* Drury. Can. Ent. 11: 217-219.

Ferris, C. D., and F. M. Brown. 1981. Butterflies of the Rocky Mountain states. Oklahoma Univ. Press, Norman.

Glassberg, J. 1993. Butterflies through binoculars. Oxford Univ. Press, Oxford, U.K.

Hammond, P. C. 1987. Ecological investigation on *Viola adunca*. U.S. Forest Serv., Siuslaw Natl. Forest, Corvallis, OR.

Hammond, P. C. 1988. Ecological investigation on *Viola adunca*. U.S. Forest Serv., Siuslaw Natl. Forest, Corvallis, OR.

Hammond, P. C. 1989. 1990-1996 management plans for the Oregon silverspot butterfly. U.S. Forest Serv., Siuslaw Natl. Forest, Corvallis, OR.

Hammond, P. C., and D. V. McCorkle. 1991. 1991 introduction of the Oregon silverspot butterfly (*Speyeria zerene hippolyta*) on Fairview Mountain. U.S. Forest Serv., Siuslaw Natl. Forest, Corvallis, OR.

Holt, R. D., and J. H. Lawton. 1994. The ecological consequences of shared natural enemies. Ann. Rev. Ecol. Syst. 25: 495-520.

Howe, W. H. 1975. The butterflies of North America. Doubleday & Co., Garden City, NY.

Iftner, D. C., J. A. Shuey, and J. V. Calhoun. 1992. Butterflies and skippers of Ohio. Ohio State Univ., Columbus.

Jones, F. M., and C. P. Kimball. 1943. The Lepidoptera of Nantucket and Martha's Vineyard. Nantucket Maria Mitchell Assoc., Nantucket, MA.

Klots, A. B. 1951. A field guide to the butterflies of North America, east of the great plains. Houghton Mifflin, Boston.

Labine, P. A. 1968. The population biology of the butterfly *Euphydryas editha*. VIII. Oviposition and its relation to patterns of oviposition in other butterflies. Evolution 22: 799-805.

Mattoon, S. O., R. D. Davis, and O. D. Spencer. 1971. Rearing techniques for species of *Speyeria* (Nymphalidae). J. Lepid. Soc. 25: 247-256.

Maynard, C. J. 1886. The butterflies of New England. Bradlie Whitten, Boston.

Nagel, H. G., T. Nightengale, and N. Dankert. 1991. Regal fritillary population estimation and natural history on Rowe Sanctuary, Nebraska. Prairie Nat. 23: 145-152.

Opler, P. A. 1983. County atlas of eastern United States butterflies (1840-1982). U.S. Fish Wildl. Serv., Washington, D.C.

Opler, P. A. 1992. A field guide to eastern butterflies. Houghton Mifflin, Boston.

Schildkneckt, C. E. 1986. Eastern regal fritillary (*S. idalia*). News. Lepid. Soc. Sept./Oct. 1986.

Schweitzer, D. F. 1987. Element stewardship abstract for *Speyeria idalia*. Report to The Nature Conservancy, Boston, MA.

Schweitzer, D. F. 1992. *Speyeria idalia*, the regal fritillary. Results of a global status survey. Report to U.S. Fish Wildlife Serv., Concord, NH.

Schweitzer, D. F. 1993. Regal fritillaries in the east. Am. Butterfl. Feb., p. 9.

Scott, J. A. 1986. The butterflies of North America. Stanford Univ. Press, Stanford, CA.

Scudder, S. H. 1889. The butterflies of the eastern United States and Canada with special reference to New England. S. H. Scudder, Cambridge, MA.

Swengel, A. 1993. Regal fritillaries: prairie royalty. Am. Butterfl. Feb., pp. 4-9.

Weed, C. M. 1926. Butterflies worth knowing. Doubleday Page and Co., Garden City, NY.

David L. Wagner: Department of Ecology and Evolutionary Biology, U-Box 43, University of Connecticut, Storrs, CT 06269-3043.

Matthew S. Wallace: Department of Ecology and Evolutionary Biology, U-Box 43, University of Connecticut, Storrs, CT 06269-3043.

George H. Boettner: Department of Entomology, Fernald Hall, University of Massachusetts, Amherst, MA 01003.

Regal Fritillary Butterfly in New England

Joseph S. Elkinton: Department of Entomology, Fernald Hall, University of Massachusetts, Amherst, MA 01003.

Index

Abagrotis crumbi benjamini, 83, 219, 221, 235
Abagrotis placida, 257
Abies, 27, 54
Abies balsamea, 27, 54
Abundance, 223
Acanaloniidae, 243, 244, 245
Acer, 5, 25
Acer rubrum, 27
Achillea millefolium, 105, 111
Acid precipitation, 71
Acorns, 128
Acrididae, 242, 244, 253
Acrobasis betulella, 255
Acrobasis comptoniella, 255
Acronicta albarufa, 83, 233
Acronicta lanceolaria, 235
Actias luna, 256
Adirondack Mountains, NY, 125
Aegeriidae, 255
Aeshnidae, 252
Agalinis, sandplain, 6, 18, 20, 83, 86, 129
Agalinis acuta, 6, 18, 20, 83, 86, 129
Agawam, MA, 198, 199
Agriculture, 1, 121, 123, 227
 in Britain, 70
 in Cutler, Maine, 59
 and forests, 6
 on Martha's Vineyard, 219
 in Massachusetts, 189, 195
 at Pineo Pond, 41
 and regal fritillary, 270
 in Vermont, 201, 208
Agriphila vulgivagella, 255
Agromyzidae, 244
Agropyron repens, 105

Agrostis, 88, 92, 105, 111
Agrostis hyemalis, 105
Airports, 7, 15, 80, 99-118, 131, 192, 193, 194, 195, 198, 199, 211-215
 as nesting areas, 191
Albumin, 264, 267, 272
Alder, 32, 41, 44, 55, 56, 139
Alder Flycatcher, 139, 141
Alexander, heart-leaved, 20
Alfalfa, 174
Alfalfa butterfly, 256
Allium vineale, 105
Alnus, 32, 55, 139
Altica sylvia, 255
Ambrosia, 40, 60
Ambrosia artemisiifolia, 105
Amelanchier, 6
Amelanchier nantucketensis, 83
American beech, 25, 27, 30, 39, 43, 48
American copper, 256
American Kestrel, 203, 204, 207, 208, 211
Amherst, MA, 199
Ammodramus henslowii, 120
 in New York, 171-186
 in Vermont, 202
Ammodramus henslowii susurrans, 6, 129
Ammodramus savannarum, 7, 86, 100, 120, 137, 141,153-170, 251
 in Massachusetts, 187-199
 in New York, 171-186, 211
 in Vermont, 202
 See also Grasshopper Sparrow
Ammodramus savannarum floridanus, 174
Amphipyra tragopoginis, 257

Index

Anagrapha falcifera, 257
Anaphalis margaritacea, 105
Anax junius, 252
Andropogon gerardii, 17, 124
Anisota stigma, 234
Antlions, 244, 253
Ants, 243, 245
Apamea burgessi, 235
Apamea mixta, 221, 236
Apharetra purpurea, 233
Aphidae, 244, 245
Aphididae, 243
Aphids, 243, 244
Aphonus tridentatus, 254
Aphrodite fritillary, 220
Apidae, 258
Apis mellifera, 258
Arachnida, 240, 244, 246
Arctiidae, 233, 234, 257
Arctostaphylos uva-ursi, 70, 88, 92
Arge, 257
Argidae, 257
Aristida purpurascens, 83
Aronia, 139, 239
Aronia arbutifolia, 88, 92
Arphia sulphurea, 253
Arphia xanthoptera, 253
Artemesia vulgaris, 104, 105, 117, 118
Arthropods, 222, 223
Asclepias, 272
Asclepias amplexicaulis, 17
Asclepias syriaca, 105, 109
Asclepias tuberosa, 5, 18, 20
Ash, 34
Asilidae, 257
Asio flammeus, 83, 86, 137, 206, 211
 See also Short-eared Owl
Aspen, 56, 65
Aster, 6, 17, 18, 20, 60, 83, 86, 88, 91, 92, 105, 129
Aster concolor, 18, 20, 83, 86
Aster dumosus, 88, 91, 92
Aster ericoides, 105
Aster linariifolius, 17

Aster novae-angliae, 17
Aster paternus, 17, 88, 92
Aster pilosus, 17
Athous rufifrons, 254
Attleboro, MA, 199
Augochlorella striata, 258
Avena, 105
Ayer, MA, 192, 193, 194, 198, 199
Bagisara rectifascia, 236
Baldfaced hornet, 258
Balsam fir, 27, 30, 54, 59
Baptista tinctoria, 124
Barn Owl, 211
Barrens, 20, 221, 228, 233-234
 blueberry, 2, 56, 62, 128, 130
 delta grassland, 27
 grassland pine, 3
 native grassland, 8, 139, 146
 pine grassland, 25-52
Barrens daggermoth, 83
Barrier beaches, 222
Bartramia longicauda, 120
 in Maine, 137, 141
 in Massachusetts, 86, 187-199
 in New York, 211
 in Vermont, 203
 See also Upland Sandpiper
Baskettail, common, 252
Bayberry, 85, 88, 90, 92, 95, 96, 105, 106, 111, 113, 117, 118, 128
Bearberry, 70, 71, 88, 92
Beavers, 125, 130
Bedford, MA, 193, 198, 199
Beech, 25, 27, 30, 39, 43, 48
Beetles, 243, 244, 255
Berkshire Mountains, MA, 188, 191
Betula, 25, 59, 139
Betula lutea, 33
Betula papyifera, 27
Betula pendula, 73
Betula populifolia, 27, 105, 109
Betula pubescens, 73
Big bluestem, 17, 124
Big-headed flies, 244
Big poplar sphinx, 257
Biogeography, 15-23

Index

Biomass, 222, 237, 238, 239, 241, 242, 243
Birch, 25, 27, 30, 33, 38, 39, 40, 41, 44, 59, 73, 105, 139
Birch tubemaker, 255
Birdfoot violet, 17, 124, 262
Birds, 137-152, 213
  breeding areas, 8, 99, 100, 137-152
  conservation, 8, 119-136, 138
  eastern grassland species, 126
  ecology, 145
  endangered, 15, 119, 148
  habitats, 99-118, 131
  Neotropical migratory, 119
  rare, 86, 99-118
  See also Specific names i.e. Sparrow
Birds, grassland, 2, 4, 7, 8, 120, 126
  abundance, 143
  and agriculture, 130
  conservation of, 130, 148
  habitats, 131, 137-152
  in Massachusetts, 187-199
  in New York, 171-186, 211-215
  interstate planning, 164
  in Vermont, 201-210
  management planning, 131, 148
  nesting, 114, 196
  origin, 127
Birds, nesting, 114, 187-199
Biting midges, 244
Bittersweet nightshade, 105
Blackberry, 104, 105, 106, 107, 109, 117, 118
Black-billed Magpie, 126
Black cherry, 105, 109, 111, 117, 118
Black huckleberry, 85, 88, 90, 92, 94, 95, 96
Black oak, 82
Black racer, 251
Black spruce, 27
Black swallowtail, 256
Blastobasidae, 255
Blastobasis, 255
Blattellidae, 253
Blazing star, 129
  See also Northern blazing star
Blinded sphinx, 257
Block Island, RI, 15, 220
  regal fritillary, 262, 263, 264
Blueberries, 27, 29, 46

  See also Barrens; Blueberry production, commercial; Lowbush blueberry
Blueberry flea beetle, 255
Blueberry production, commercial, 8, 25, 26, 27, 35-36, 139, 146, 237, 238, 239
Blue-eyed grass. *See* Sandplain blue-eyed grass
Bluejoint reedgrass, 2, 53, 56, 64
Bluejoint reedgrass grasslands, 53-67
Blue List, 172
Blue toadflax, 105, 111
Bluets, 252
Blue violet, common, 262
Blunt-leaved milkweed, 17
Bobolink, 6, 127, 129
  in Maine, 141, 143, 145, 147
  in Massachusetts, 187-199
  in New York, 171, 175, 179, 180, 182
  in Vermont, 202, 204, 207, 208
Boloria bellona, 220, 256
Boston, MA, 193, 198, 199
  See also Logan Airport
Bouncing bet, 105
Bracken, 25, 27, 40, 41, 44, 48
Braconidae, 245, 248
Bradley Field, CT, 15
  See also Airports
Breeding
  birds, 176, 213
  habitats, 189
  regal fritillary, 262, 264
Breeding areas, 149, 194, 195
  in Massachusetts, 190
  of sparrows, 178, 183
  See also Nesting areas
Breeding Birds Survey, 202, 203, 204
Britain. *See* Great Britain
Broad-leaved golden aster, 18, 20
Brooklyn, NY, 99-118, 211-215
Brown-bordered geometer, 256
Brown-collared dart, 257
Brown elfin, 256
Brown-headed Cowbird, 126, 181
Brown Thrasher, 137, 141, 142, 143, 146
Brushcutting, 72
Buckeye, 256

279

Index

Buprestidae, 254
Burning, 121, 139, 148, 159, 165, 166, 167, 227, 238, 239
   annual vs. biennial, 96
   in Connecticut, 96
   at Floyd Bennett Field, 99–118
   on Nantucket Island, 85–98
   See also Fire
Burning, prescribed, 54, 82, 130, 159, 196, 226
Burn rotation, 159, 161, 167
Bush clover, 105, 106
Bushy aster, 88, 91, 92
Bushy rockrose, 6, 18, 20, 83, 86, 91, 129
Butter-and-eggs, 105
Butterflies, 219, 220, 224, 251, 256
   eggs, 268
   nutritional needs, 272
   See also Specific butterfly i.e. Regal fritillary
Butterfly weed, 5, 18, 20, 272
Calamagrostis canadensis, 2, 53
Calluna vulgaris, 70
Calophasia lunula, 257
Calosoma calidum, 254
Calosoma sycophanta, 254
Camp Edwards Military Reservation, MA, 192, 193
Canada goldenrod, 105
Cape Cod, MA, 7
   birds, 193, 194
   grasslands, 8, 188
   heathlands, 70, 72, 86
   moths, 221
   sandplain grasslands, 86
   sparrows, 190
   See also Martha's Vineyard; Massachusetts; Nantucket Island
Carabidae, 254
Carex, 105, 109, 139, 176
Carex lucorum, 17
Carex pensylvanica, 17, 89, 93, 239
Carex pensylvanica/umbellata, 88, 92
Carmenta anthracipennis, 255
Carolina locust, 253
Castor canadensis, 125
Catocala, 234
Catocala herodias gerhardi, 83, 233
Cecropia moth, 256

Cedar, 33
Celastrina argiolus, 256
Celastrus scandens, 105
Celery looper moth, 257
Celithemis elisa, 252
Center for Biological Conservation, 187, 188, 194, 195, 196
Cerambycidae, 255
Ceratopogonidae, 244
Cercopidae, 244, 245
Cercyonis pegala alope, 256
Chaetaglaea cerata, 234
Chaetaglaea tremula, 234
Chain-dotted geometer, 227
Champlain Lowlands, VT, 201, 202, 205, 206, 208
Change, environmental, 223, 224
Charadrius vociferus, 139, 141
Charcoal analysis, 25, 27, 29, 30, 32, 56, 59, 61, 138
Chauliodes pectinicornis, 253
Chicopee, MA, 164, 192, 193, 194, 195, 198, 199
Chloropidae, 243, 244, 245
Chokeberry, 139, 239
Chondestes grammacus, 127
Chorthippus curtipennis, 253
Chrysochus auratus, 255
Chrysomelidae, 243, 244, 245, 255
Chrysopsis falcata, 18, 20, 129
   See also Pityopsis falcata
Chrysopsis mariana, 18, 20
Chytonix sensilis, 234
Cicadellidae, 244, 245
Cicindela punctulata, 254
Cicindelidae, 254
Cicinnus melsheimeri, 83, 221, 233
Cingilia catenaria, 227, 235
Circus cyaneus, 86, 120, 137, 141, 211, 227
Cirsium, 105, 272
Cisthene packardii, 233
Cistothorus platensis, 206
Civilian Conservation Corps, 80
Cleridae, 254

280

Index

Climbing bittersweet, 105
Clinton, MA, 189, 199
Clouded sulphur, 256
Clover, 17, 174
Clover hayworm moth, 255
Coastal barrens buckmoth, 83
Coastal heathland cutworm, 83
Coastal plain, 20
Coastal sandplain, 87, 217-236
Coastal sandplain grasslands
  on Nantucket Island, 85-98
Coenagrionidae, 252
Coenonympha tullia inornata, 256
Coleophora, 255
Coleophoridae, 255
Coleoptera, 244, 252, 254, 255
Colias eurytheme, 256
Colias interior, 256
Colias philodice, 256
Colinus virginianus, 126
Coluber constrictor, 251
Common baskettail, 252
Common blackberry, 105
Common blue violet, 262
Common dandelion, 105
Common elderberry, 105
Common forktail, 252
Common green darner, 252
Common milkweed, 105, 109
Common mugwort, 104, 105, 106, 109, 117, 118
Common tanwave, 256
Common wood nymph, 256
Common Yellowthroat, 137, 141, 142, 143, 146, 147
Comptonia peregrina, 25
Conanicut Island, RI, 264
Connecticut
  Bradley Field, 15
  grasslands, 96
  Hartford, 15
  North Haven Sand Plains, 15, 125, 130
  regal fritillary, 262
Connecticut River valley, MA, 188, 195
  sparrows, 190
Connecticut River valley, VT, 206, 208

Conocephalus fasciatus, 253
Conservation, 7, 227
  birds, 8, 119-136, 138, 148, 172, 182, 201-210
  of early-successional communities, 217-236
  of East Stream Grassland, 65
  grassland birds, 131
  of grasslands, 21, 238
  history, 79
  invertebrates, 249
  of species diversity, 222
Conservation Reserve Program, 131
Contopus virens, 126
Cooperative management planning, 164, 167
Copris fricator, 254
Coral hairstreak, 256
Cord grass, freshwater, 17
Corduliidae, 252
Corydalidae, 253
Corythucha cydoniae, 253
Cotalpa lanigera, 254
Crambidia pallida, 257
Crambus leachellus, 255
Creophilus maxillosus, 254
Crickets, 243, 244
Crymodes devastator, 257
Cryptocephalus gibbicollis, 255
Ctenicera sjaelandica, 254
Cucullia speyeri, 221, 235
Cudweed, purple, 18
Cut-leaved blackberry, 105
Cutler, ME, 53-67
Cutworm, 83
Cutworm moth, 219, 221
Cycnia inopinatus, 221, 234
Cymus, 253
Dairy industry, 183
Daisy fleabane, 105, 109
Damsel bugs, 244
Danaus plexippus, 256
Dandelion, common, 105
Danthonia, 139
Danthonia spicata, 17, 92, 239, 251
Dartmouth, MA, 199
Dasymutilla vesta, 257

Index

Datana major, 236
Daucus carota, 105
Decodon stemborer, 221
Deerfield, MA, 199
Deforestation, 70
Delphacidae, 244, 253
Delphacid planthoppers, 244
Delta, 27
Demographics, 166
Deschampsia flexuosa, 71
Development, 188, 217
Dewberry, 88, 99, 102, 104, 105, 106, 107, 109, 110, 111, 112, 117, 118
Dicerca, 254
Dichelonyx, 255
Dickcissel, 127
Dictyopharidae, 243, 244, 245
Dictyopharid planthopper, 242, 244
Didymops transversa, 252
Dingy cutworm moth, 257
Diplodia pinea, 81
Diptera, 243, 244, 257
Dispersal, 7, 10, 19
  invertebrates, 238
  prairie flora, 20
  rates, 155, 162, 165
Dissosteira carolina, 253
Disturbances, 81, 82, 86, 219, 226, 238, 270
  See also Fire; Weather
Diversity, 54, 246, 247
  of herbaceous species, 96
  of insects, 237-250
  of invertebrates, 238, 240, 241, 242, 248
  on Kennebunk Plains, 251
  species, 89, 221, 222, 223
Dock, 40
Dogbane beetle, 255
Dolichonyx oryzivorus, 6, 127, 141, 145, 171
  in Massachusetts, 187-199
  in Vermont, 202
  See also Bobolink
Dolichopodidae, 244
Dolichovespula maculata, 258
Downy birch, 73
Downy goldenrod, 17

Drasteria occulta, 257
Dropseed grass, 20
Dryocampa rubicunda, 256
Dumetella carolinensis, 213
Dwarf sumac, 105
Eacles imperialis, 83, 221, 234
Early goldenrod, 17, 105
Early successional
  birds, 138, 146
  communities, 217, 226
  habitats, 1, 7
  species, 40, 146
Eastern hemlock, 25, 27, 39, 40, 41, 43, 48, 59, 123
Eastern Meadowlark, 127, 175
  in Maine, 141, 143, 145, 147
  in Massachusetts, 187-199
  in New York, 175, 179, 180, 211, 213
  in Vermont, 202, 204, 207, 208
Eastern silvery aster, 18, 20, 83, 86
Eastern Wood-Pewee, 126
East Stream Grasslands, 53-67
  description, 54-56
  fire, 55, 62, 63
  history, 58-60
  management planning, 54, 65
  vegetation, 56, 63, 64
Ecology, 5, 26
  and birds, 145, 196
  East Stream Grassland, 65
  grasslands, 9
  heathlands, 9, 69
  Martha's Vineyard, 81
Ecosystems, 9, 26, 54
  management strategy, 82, 222
  Martha's Vineyard, 221
Edith's checkerspot, 268
Elateridae, 254
Elderberry, common, 105
Elisa skimmer, 252
Elizabeth Islands, MA, 188, 191, 196
  grasslands, 8
Elm, 34
Empidonax alnorum, 140, 141
Enallagma, 252
Endangered species, 21
  birds, 15, 119, 189, 194, 205, 211
  butterflies, 224
  communities, 86
  Grasshopper Sparrow, 153
  heathlands, 70

Index

Kennebunk Plains, 237
  in Maine, 153
  in Massachusetts, 187, 190
  on Nantucket Island, 93
  regal fritillary, 264
  sparrows, 172
  See also Plants
England. *See* Great Britain
Enoclerus muttkowskii, 254
Enodia anthedon, 256
Environmental changes, 223, 224, 228
Epitheca cynosura, 252
Epping Plain, ME, 3, 33, 34, 123
Eragrostis curvula, 104, 105, 117, 118
Eragrostis spectabilis, 105
Erigeron annuus, 105, 109
Eremophila alpestris, 6, 140, 141
Eremophila alpestris praticola, 127
Ericaceae, 60
Ericaceous vegetation, 5, 29, 45, 46
Erigeron annuus, 105, 109
Eucoptocnemis fimbriaris, 234
Eucosma similiana, 255
Eumacaria latiferrugata, 235, 256
Euphydryas editha, 268
European mantid, 253
Euspilotus assimilis, 254
Euthamia graminifolia, 17
Euthamia tenuifolia, 105
Euthamia tenuifolia/graminifolia, 92
Euxoa mimallonis, 257
Euxoa violaris, 235
Evolution, 5, 8
Extinction, 6, 9, 166, 223, 224, 238
  probability, 160, 161, 162
  risks, 154, 165, 167
Extirpation, 6, 70, 204, 206, 208, 219, 221, 223, 225, 238, 248, 264
  butterflies, 224
  Heath Hen, 79
  sparrows, 172
Fagus grandifolia, 25
Falco sparverius, 208, 211
Falmouth, MA, 199
Faronta rubripennis, 235
Feltia jaculifera, 257
Fern, 25, 27, 31, 41, 44, 46, 48
Fertilizer, 70

Fescue, 105
Festuca capillata, 105
Festuca elatior, 105
Festuca gigantea, 105
Festuca myuros, 105
Festuca ovina, 85, 88, 92, 105, 109, 251
Festuca rubra, 104, 105, 117, 118
Field garlic, 105
Field sorrel, 105
Field Sparrow, 137, 141, 146, 147
Finger Lakes National Forest, NY, 171-186
Fir, 27, 30, 54, 56, 59
Fire, 4, 5, 8, 21, 48, 62, 79, 81, 122, 123, 188, 218, 219, 222, 223, 226
  causes, 47
  East Stream Grassland, 53, 54, 55, 58, 63
  and ecosystems, 26
  at Floyd Bennett Field, 99-118
  Great Miramachi, 34
  history, 27
  as management technique, 72
  Pineo Ridge, 25-53
  prescribed, 72, 99-118, 213, 248
  suppression, 70
  and vegetation, 46
  See also Burning; Disturbances
Firebreaks, 80, 82
Fire rotation period, 26
Fire-stops. *See* Firebreaks
Flax, sandplain, 18, 20, 83, 86, 91
Flies, 244
Flora, prairie, 18, 20
Floyd Bennett Field, NY, 99-118, 211-215
  species list, 105
Forbs, 89, 94, 139, 180, 239
  in Ram Pasture, 89, 90
Forests, 1, 70
  history, 6, 126, 138
  in Massachusetts, 189
  on Martha's Vineyard, 81
Forktail, common, 252
Formicidae, 243, 245, 258
Fort Devens, MA, 192, 193, 194
Frankia, 46
Fraxinus, 34
Fresh-water cord-grass, 17
Fritillary butterfly, 224
  See also Regal fritillary

283

Index

Frost bottom, 224, 225, 227, 228, 233, 234
Frostweed, 17
Fruit flies, 243, 244
Fungus, 81
Fuscatum, 18
Game birds, 128
Gamochaeta purpurea, 18
Gardner, MA, 198, 199
Gateway National Recreation Area, 100
  See also Brooklyn, NY; Floyd Bennett Field
Gaylussacia, 128
Gaylussacia baccata, 85, 88, 92, 94, 95
Gelechiidae, 255
Genetic Analysis, 10
Geometridae, 233, 235, 236, 256
Geothlypis trichas, 137, 141
Geotrupes splendidus, 254
Gerhard's underwing, 83
Giant fescue, 105
Glaciers, 6, 15, 16, 21, 126, 218, 224
Glassy cutworm moth, 257
Glena cognataria, 235
Glyphosate. *See* Roundup
Gnaphalium obtusifolium, 105, 111
Goat's rue, 124
Goldenrod, 17, 20, 92, 105, 106, 109, 111, 171, 172, 176, 178, 180
Goldsmith beetle, 254
Gomphidae, 252
Gomphus exilis, 252
Gramineae, 40, 41
Graminoids, 5, 87, 94, 105, 139, 145, 239
  in Ram Pasture, 89, 90
Grammia phyllira, 234
Grass, 17, 20, 60, 102, 105, 111
  as breeding areas, 180
  rhizomatous, 6
Grasshoppers, 244, 253
Grasshopper Sparrow, 7, 8, 86, 120, 127, 129, 131
  in Maine, 137, 141, 145, 147, 148, 149, 153-170, 251
  management planning, 153
  in Massachusetts, 187-199
  in New York, 100, 171-186, 211, 212
  species decline, 213
  in Vermont, 202, 203, 204, 206, 207
Grassland barrens, 139, 142, 146, 148
  in Maine, 140, 149
  See also Blueberry barrens
Grassland birds
  and agriculture, 130
  conservation of, 130, 148
  habitats, 99-118, 131, 137-152, 211-215
  in Maine, 137-152
  and management planning, 131, 148, 164, 187
  in Massachusetts, 187-199
  nesting, 7, 114, 196
  in New York, 99-118, 171-186, 211-215
  origin, 127
  in Vermont, 201-210
  See also Specific bird i.e. Henslow's Sparrow
Grassland obligates, 228
Grasslands, 2
  agricultural, 3
  artificial, 189
  and balance with forests, 65
  as bird habitats, 99-118, 137, 138
  birds, 8, 113, 120
  Connecticut, 96
  conservation, 8
  decline of, 138, 220
  ecology, 9
  endemic plants, 6
  habitats, 248
  history, 6, 7, 16-20, 27, 119-136, 120, 188
  insects, 2, 262
  maintenance, 113
  management planning, 8, 21, 53, 196
  in Massachusetts, 188
  midwestern, 19
  migration of flora, 22
  native, 2, 8
  New England, 15-23
  paleoecology, 3
  plants 6, 15-23
  preservation of, 184
  research needs, 9
  taxa, 21
  See also East Stream Grasslands
Grasslands, bluejoint reedgrass, 53-67
Grasslands, managed, 174
Grasslands, maritime, 217-236
Grasslands, sandplain, 238, 251
  and regal fritillary, 261

Index

Grass-leaved goldenrod, 17
Grass-leaved ladies' tresses, 83
Gray birch, 27, 105, 109, 111
Gray Catbird, 213, 214
Gray goldenrod, 17
Grazing, 70, 85, 124, 171-186, 182, 219, 224
    as conservation tool, 66
    effects on sparrows, 175
    rotational, 75
Great Britain, 69-78
Greater Prairie-Chicken, 2, 79, 128, 189
    See also Heath Hen
Great golden digger, 258
Great Miramachi Fire, 34
Great spangled fritillary, 256
Green darner, common, 252
Green metallic bee, 243, 245
Grubbing out, 73
Gryllidae, 243, 244, 245, 253
Gryllus, 253
Gunthion. *See* Insecticides
Gypsy moth, 257
Habitat obligates, 222, 223
Habitats, 6
    birds in Maine, 137-152
    birds in New York, 99-118, 171-186, 211-215
    birds in Vermont, 201
    breeding, 189, 194
    endangered species, 222, 237
    of Grasshopper Sparrows, 153
    of grassland birds, 131
    heterogeneity, 226
    loss of, 188
    and management planning, 166, 213, 227
    Martha's Vineyard, 219
    modification, 155
    moths, 233-236
    regal fritillary, 262, 263, 271
    requirements for birds, 188
    restoration of, 211-215
    size, 149
    sparrows, 171-186, 178
Habitats, grassland, 99-118, 171-186, 238, 248
Hadley, MA, 193, 198
Hairgrass, 71
Hair-like fescue, 105

Hairy bush-clover, 17
Hairy wild lettuce, 83
Halictidae, 243, 245, 258
Hanscom Field, MA, 193
Harpalus, 254
Hartford, CT, 15
Hatfield, MA, 193, 198
Hawley, MA, 199
Hawthorn lacebug, 253
Hayfields. *See* Pastures
Heart-leaved alexander, 20
Heath, 60
Heath aster, 17
Heather, 70, 71
Heath Hen, 2, 4, 5, 6, 7, 79, 127, 128, 129, 189
    See also Tympanuchus cupido cupido
Heathlands, 2, 60, 69-78, 86, 218, 227, 235, 236
    and brushcutting, 72
    ecology, 9, 69-78
    history, 6, 7
    maintenance, 75
    management planning, 71, 73
    Nantucket Island, 75, 123, 124
    nutrient levels, 71
    regal fritillary, 261
    restoration of, 69-78
    vegetation, 7, 71, 76, 85
Hector Land Use Area. *See* Finger Lakes National Forest
Helianthemum canadense, 17
Helianthemum dumosum, 6, 18, 20, 83, 86, 129
Helianthemum propinquum, 18, 20
Hemaris diffinis, 257
Hemaris gracilis, 227, 236
Hemileuca maia, 234
Hemileuca maia maia, 83
Hemipachnobia monochromatea, 236
Hemiptera, 244, 245, 253
Hemlock, 25, 27, 33, 34, 39, 40, 41, 43, 44, 48, 59, 123
Hempstead Plains, NY, 3, 15, 21, 124, 125, 130
    Heath Hen, 129
    native grasslands, 8
    See also Long Island
Henslow's Sparrow, 6, 120, 129, 131

Index

in New York, 171-186
in Vermont, 202, 203, 204, 205, 206, 207
Herbaceous species, 96
Herbicides, 8, 27, 74, 85, 96, 139, 145, 148, 155, 156, 238
See also Insecticides
Herpetogramma thestealis, 255
Hesperoleon abdominalis, 253
Heterogeneity, habitat, 226
Highbush blueberry, 27
Hippocrepis comosa, 224
Histeridae, 254
Holometabolism, 223
Homoptera, 243, 244, 245, 253
Honeybee, 258
Horned Lark, 6, 127, 131, 140, 141, 203, 204, 207
Huckleberry, 85, 90, 128
Hyalophora cecropia, 256
Hydraecia immanis, 235
Hydrochara, 254
Hydrophilidae, 254
Hymenoptera, 243, 245, 257, 258
Hypatopa, 255
Hypocoena defecta, 236
Hypsithermal period, 4, 18
Hypsopygia costalis, 255
Ichneumon eximus, 257
Ichneumonidae, 245, 248, 257
Imperial moth, 83, 221, 225
Incisalia augustinus, 256
Indian grass, 17
Indigo, 124
Indigo Bunting, 140, 141
Inornate ringlet, 256
Insecticides, 8, 237-250
See also Herbicides
Insects, 217-236, 237-250
conservation, 217-236
diversity, 237-250
endemic, 6
grasslands, 9, 262
of Kennebunk Plains, 251-260
Instar, 261, 262, 269, 270
Invasive vegetation, 73
Invertebrates, 9, 16, 238

conservation of, 249
diversity, 240, 241, 242, 248
Ipswich Savannah Sparrow, 6
Ironweed, 272
Ischnura verticallis, 252
Isophrictis, 255
Itame, 233
Jack pine, 26, 37, 39, 81
Jamaica Bay Wildlife Refuge. *See* Floyd Bennett Field
Japanese knotweed, 104, 105, 106, 109, 111
Jeopardized species, 148
See also Endangered species
Juncus effusus, 105, 109
Juncus greenei, 17
Juniperus virginiana, 105
Junonia coenia, 256
Kalmia angustifolia, 27
Katama Plain, MA, 227, 263
See also Martha's Vineyard
Kennebunk, ME, 155, 156, 157, 159, 160, 163, 164, 165
Kennebunk Plains, 9, 15, 237-250, 251-260
See also York County
Kestrel, American, 203, 204, 207, 208, 211
Killdeer, 139, 141
Lace bugs, 244
Lacinipolia meditata, 257
Lactuca canadensis, 105, 111
Lactuca hirsuta, 83
Lactuca scariola, 105
Lady's thumb, 105
Lake Memphramagog, VT, 206
Lampyridae, 254
Lance-leaved goldenrod, 105
Lance-leaved violet, 262
Landfills, 189, 194, 199
Lanius ludovicianus, 120, 205
Larch, 34
Large maple spanworm moth, 256
Large yellow underwing, 257
Larix, 34
Lark Sparrow, 127

Index

Late goldenrod, 17
Laurel, 27, 46
Lauxaniidae, 244, 245
Lauxaniid flies, 244
Leaf beetles, 243, 244
Leaf bugs, 244
Leafhoppers, 244
Leaf miner flies, 244
Lebia viridis, 254
Lepidoptera, 243, 244, 255, 256, 257, 261-275
Lepidopterans, 217-236
   See also Insects
Lepipolys perscripta, 235
Lespedeza, 174
Lespedeza capitata, 17
Lespedeza cuneata, 105
Lespedeza hirta, 17
Lettuce, 105
Leucorrhinia hudsonicus, 252
Leverett, MA, 198
Levittown, NY, 3
Liatris scariosa , 4, 18, 20, 129, 139, 239, 251
   See also Northern blazing star
Libellula julia, 252
Libellula lydia, 252
Libellula pulchella, 252
Libellulidae, 252
Liburniella ornata, 253
Lichen, 105
Lightning, 47
Limenitis archippus , 256
Limenitis arthemis, 256
Limenitis arthemis astyanax, 256
Linaria canadensis, 105, 111
Linaria vulgaris, 105
Lincoln, MA, 195
Lincoln's Sparrow, 140, 141
Linum intercursum, 18, 20, 83, 86
Lithophane thaxteri, 234
Litter cover, 71, 99, 100, 104, 105, 106, 107, 108, 109, 111
Little bluestem, 5, 17, 124, 251
   in Maine, 139, 239
   on Nantucket Island, 85, 88, 89, 90, 92, 94, 96
   in New York, 16, 99, 102, 104, 105, 107, 109, 110, 111, 112, 113, 117, 118
Little wood satyr, 256
Loblolly pine, 81
Logan Airport, MA, 193
   See also Airports
Loggerhead Shrike, 120, 204, 205, 206, 207
Logging, 32, 45, 58, 60, 70
Long-horned grasshopper, 244, 253
Long Island, NY, 10, 16
   Heath Hens, 129
   Hempstead Plains, 3, 15, 124, 125, 130
   Montauk Downs, 125
   native grasslands, 8
   regal fritillary, 264
   See also Hempstead Plains; New York
Long-legged flies, 244
Love grass, 104, 105, 106, 117, 118
Lowbush blueberry, 8, 25, 27, 35, 36, 54, 56, 88, 90, 92, 95, 123, 139, 237, 239, 251
Low rockrose, 18, 20
Lucidota nigricans, 254
Lumber industry. *See* Logging
Luna moth, 256
Lupine, wild, 18, 20
Lupinus perennis, 18, 20
Lycaena phlaeas americana, 256
Lycaenidae, 256
Lycia ypsilon, 221, 233
Lygaeidae, 243, 244, 245, 253
Lymantria dispar, 257
Lymantriidae, 257
Lysandra bellargus, 224
Lysimachia quadrifolia, 139, 239
Macrodactylus subspinosus, 255
Macromiidae, 252
Maine, 4
   agriculture, 33, 59
   birds, 130, 143, 184
   blueberry barrens, 130
   conservation, 8, 130
   Cutler, 53-67
   East Stream Grassland, 53
   ecosystems, 54, 138

Index

Epping Plain, 33, 34
Grasshopper Sparrow, 8, 153-170
grassland birds, 137-152
grassland habitats, 248
Kennebunk, 155, 156, 157, 159, 160, 163, 164, 165
Kennebunk Plains, 15, 237-250, 251-260
logging, 32
Mud Pond, 25-52
Narraguagus River, 29
native grasslands, 8
Pineo Ridge, 25-52
sandplain grasslands, 238
Sanford, 155, 156, 157, 158, 159, 160, 163, 165
vegetation changes, 43
Washington County, 3, 25-52, 54
Wells, 155, 156, 157, 158, 159, 160, 163, 165
York County, 9, 239

Management
  of ecosystems, 221, 222

Management cooperation, 167

Management planning, 160, 165
  bird habitats, 99-118, 166, 213
  early-successional communities, 217-236
  East Stream Grassland, 65
  for Grasshopper Sparrows, 153, 164
  for grassland birds, 131, 148, 187
  grasslands, 21, 53, 81, 95, 96, 166
  grubbing out, 73
  habitats, 159, 227, 248
  heathlands, 69, 70, 71
  Martha's Vineyard, 82
  on Nantucket Island, 85, 88, 89, 90, 92, 94, 96
  nesting sites, 196
  pastures, 171, 182
  prescribed fires, 72
  regal fritillary, 261, 264, 271
  sandplains, 221
  using population viability analysis, 154
  vegetation on Nantucket Island, 85-98

Mantidae, 253

Mantis religiosa, 253

Manuel F. Correllus State Forest.
  See Martha's Vineyard

Many-flowered aster, 105

Maple, 5, 25, 27, 30, 34, 41

Maritime grasslands, 217-236

Maritime spruce, 54

Martha's Vineyard, MA, 4, 7, 15, 16, 79-84, 188

birds, 190, 194
ecology, 81
ecosystems, 221
forests, 81
grasslands, 8, 188
Heath Hens, 128
heathlands, 70, 72
insects, 217-236
management planning, 81, 82
Manuel F. Correllus State Forest, 79, 80, 83
moths, 233-236
nesting areas, 191
rare species, 83
regal fritillary, 263, 264
sandplain grasslands, 79-84
sandplain management, 217-236
sparrows, 199
vegetation, 82
See also Cape Cod, MA; Massachusetts; Nantucket Island, MA

Massachusetts
  Agawam, 198, 199
  Amherst, 199
  Attleboro, 199
  Ayer, 192, 193, 194, 198, 199
  Bedford, 193, 198, 199
  Berkshire Mountains, 188, 191
  Boston, 193, 198, 199
  Camp Edwards Military Reservation, 192, 193
  Cape Cod, 7, 8, 70, 72, 86, 188, 190, 193, 194, 221
  Chicopee, 164, 192-196, 198, 199
  Clinton, 189, 199
  Connecticut River Valley, 188, 190, 195
  Dartmouth, 199
  Deerfield, 199
  Elizabeth Islands, 8, 188, 191
  Falmouth, 199
  Fort Devens, 192, 193, 194
  Gardner, 198, 199
  grassland areas, 188
  grassland birds, 187-199
  Hadley, 193, 198
  Hanscom Field, 193
  Hatfield, 193, 198
  Hawley, 199
  Leverett, 198
  Lincoln, 195
  Logan Airport, 193
  Martha's Vineyard, 7, 8, 15, 16, 70, 72, 79-84, 86, 128, 188, 190, 194, 199, 217-236, 262, 263, 264
  Middleboro, 198

Index

Nantucket Island, 7, 8, 15, 16, 70, 72, 75, 85-98, 124, 188, 190, 196, 233-236, 263, 264
Nashawena Island, 191, 194, 199
Naushon Island, 199, 219, 264
Newburyport, 198
No Mans Land, 220, 264
Nonamesset Island, 199
Northampton, 193, 199
Orange, 193, 198, 199
Otis Air Force Base, 192
Pasque Island, 199
Pepperell, 198
Plainfield, 199
Plymouth, 193, 198, 199
Provincetown, 191, 193, 198
Quabbin Reservation, 125
regal fritillary, 261, 262
Sandwich, 192, 193, 198, 199
Sheffield, 199
Springfield, 15
Stow, 199
Sunderland, 198, 199
Truro, 193, 198
Turners Falls, 194, 198, 199
Wellfleet, 193, 198, 199
W. Bridgewater, 199
Westfield, 198, 199
Westover Air Force Base, 15, 164, 191-196
Worcester, 189, 194, 199
Worthington, 199

Massachusetts Audubon Society, 81, 85-97, 187, 196, 264

Massachusetts Division of Fisheries and Wildlife, 187

Massachusetts Natural Heritage and Endangered Species Program, 91, 196

Meadow fescue, 105

Meadow fritillary, 220, 256

Medicago sativa, 174

Megisto cymela, 256

Melanerpes carolinus, 126

Melanerpes erythrocephalus, 126

Melanoplus bivittatus, 253

Melanoplus confusus, 253

Melanoplus fasciatus, 253

Melanoplus f. femurrubrum, 253

Melanoplus keeleri luridus, 253

Melanoplus s. sanguinipes, 253

Melospiza lincolnii, 140, 141

Melospiza melodia, 137, 141

Melsheimer's sack bearer, 83, 221

Metalectra richardsi, 234

Metapopulation analysis, 9, 10

Metarranthis, 221

Metarranthis apiciaria, 221, 233

Metarranthis pilosaria, 236

Middleboro, MA, 198

Migration
  grassland flora, 22
  plant, 18, 20

Milkweed, 17, 105, 109, 272

Miller moths, 244

Mimallonidae, 233

Minnesota Boundary Waters, 26

Miridae, 244

Molinia caerulea, 71

Molothrus ater, 126, 181

Monarch, 256

Montauk Downs, NY, 125

Mordellidae, 244

Mormon fritillary, 269, 272

Morrisonia mucens, 234

Moss, 105, 109, 111

Moths, 83, 221, 224, 227, 233-236, 248, 256, 257

Mourning cloak, 256

Mowing, 8, 99, 102, 106, 113, 114, 139, 165, 167, 181, 182, 189, 196, 213, 238, 239
  on Nantucket Island, 85-98

Mud Pond, 25, 29, 30, 123
  chronology, 32
  fires, 43, 48
  history, 36-41
  pollen, 42
  vegetation, 46
  See also Pineo Ridge

Mugwort, common, 104, 105, 106, 109, 117, 118

Mullein, common, 105

Multiflora rose, 105

Muscidae, 244, 245

Muscid flies, 244

Mutillidae, 257

Myrica, 31

Myricaceae, 31

Myrica pensylvanica, 85, 88, 92, 95, 105, 111, 117, 118, 128

289

Index

Myrmeleontidae, 244, 253
Nabidae, 244, 245
Nadata gibbosa, 257
Nantucket Island, MA, 4, 7, 15, 16
   birds, 190, 196
   burning, 114
   deforestation, 70, 123, 124
   grasslands, 8, 188
   heathlands, 15, 70, 75, 124
   lepidoterans, 219
   managing grasslands, 96
   moths, 233-236
   nesting areas, 191
   prescribed fires, 72
   regal fritillary, 262, 263, 264
   vegetation management, 85-98
   See also Cape Cod; Martha's Vineyard; Massachusetts
Nantucket shadbush, 83
Narraguagus River, ME, 29
Nashawena Island, MA, 194, 199
   See also Elizabeth Islands
Native Americans, 4, 21, 70, 121, 122, 188, 219
   agriculture, 123
Nature Conservancy, 238, 252, 264
Naushon Island, MA, 199, 219
   regal fritillary, 264
Necrodes surinamensis, 254
Necrophila americana, 254
Nectar, 261, 264, 271, 272
Nehalennia irene, 252
Neotropical migrant birds, 138
Nesting, 114, 123
Nesting areas, 191, 204, 205
   See also Breeding areas
Nesting birds, 187-199
Neuroptera, 244, 253
Newburyport, MA, 198
New England aster, 17
New York
   Adirondack Mountains, 125
   Brooklyn, 99-118, 211-215
   endangered species, 211
   Finger Lakes National Forest, 171-186
   Floyd Bennett Field, 99-118, 211-215
   Gateway National Recreation Area, 100, 211
   grassland birds, 171-186, 211-215
   grazing lands, 171-186

Hempstead Plains, 3, 8, 15, 21, 124, 125, 129, 130
Jamaica Bay Wildlife Refuge, 211
Levittown, 3
Long Island, 8, 10, 15, 16, 21, 129, 130, 264
Nicrophorus defodiens, 254
Nicrophorus marginatus, 254
Nicrophorus orbicollis, 254
Nicrophorus tomentosus, 254
Nitrogen, 71
Noctua pronuba, 221, 257
Noctuid, 221
Noctuidae, 233, 234, 235, 236, 257
Nola sorghiella, 235
No Mans Land, MA, 220, 264
Nomotettix c. cristatus, 253
Nonamesset Island, MA, 199
Northampton, MA, 193, 199
Northeastern spruce fir, 27
Northeast Highlands, VT, 208
Northern blazing star, 4, 5, 6, 18, 20, 139, 239, 251
Northern Bobwhite, 126
Northern Harrier, 86, 120, 137, 139, 141, 148, 149, 204, 207, 211, 213, 227
Northern pearly eye, 256
Northern petrophora, 256
Northern red oak, 25, 27, 29
Northern Spotted Owl, 154
Northern wood roach, 253
North Haven Sand Plains, CT, 15, 125, 130
Notodontidae, 236, 257
Nova Scotia, 6
Nuclear polyhedrosis virus (NPV), 271
Nutrients
   heathlands, 71, 74, 75
Nutritional needs
   lepidopterans, 225
   regal fritillary, 272
Nuttall's milkwort, 83
Nymphalidae, 256, 261-275
Nymphalis antiopa, 256
Oak, 5, 25, 27, 29, 30, 34, 81, 82, 83, 124, 227
Oat, 105
Obligates, 220, 222, 223, 226, 227, 228

Index

Occult drasteria, 257
Ochrocephala, 255
Odonata, 252
Oecanthidae, 253
Oecanthus quadripunctatus, 253
Oiceoptoma noveboracense, 254
Olethreutes cespitana, 255
One-eyed sphinx, 257
Ontholestes cingulatus, 254
Onthophagus hecate, 254
Onthophagus nuchicornis, 254
Orange, MA, 193, 198, 199
Oregon silverspot, 271
Orphulella speciosa, 253
Orthoptera, 244, 253
Oryzopsis pungens, 17
Otis Air Force Base, MA, 192
Ovate-leaved violet, 17, 124, 262
Owls, 83, 86, 137, 148, 206, 211
Oxalis stricta, 105, 111
Pachybrachys, 255
Pachysphinx modesta, 257
Painted lady, 256
Paleoecology, 3, 29, 48, 53
Panic grass, 105, 106
Panicum, 105
Panton, VT, 205
Papaipema appassionata, 236
Papaipema harrisii, 236
Papaipema stenocelis, 236
Papaipema sulphurata, 221, 236
Paper birch, 27
  See also White birch
Paper wasp, 258
Papilio canadensis, 256
Papilionidae, 256
Papilio polyxenes, 256
Papillose nutsedge, 83, 93
Parasitic wasps, 245
Parcoblatta virginica, 253
Parthenocissus quinquefolia, 105
Pasque Island, MA, 199
Passerculus sandwichensis, 6, 137, 141, 171, 208
  in Massachusetts, 187-199
  in New York, 213
  See also Savannah Sparrow
Passerculus sandwichensis princeps, 6
Passerina cyanea, 140, 141
Pastures, 6, 7, 142, 148, 149, 270
  management planning, 171, 182
  in New York, 171-186
Pearly everlasting, 105
Peat, 70
Peatlands, 27
Pelecinidae, 257
Pelecinus polyturator, 257
Pennsylvania, 263
Pennsylvania sedge, 89, 90, 93, 239
Pentatomidae, 244
Pepperell, MA, 198
Pesticides. *See* Insecticides
Petrophora subaequaria, 256
Phalonia hospes, 248, 255
Phaneta, 255
Phlegyas abbreviatus, 253
Phleum pratense, 3
Phosphorus, 72
Photinia, 139, 239
Photinia floribunda, 88, 92
Phragmites communis, 105
Phyllobaenus pallipennis, 254
Phyllophaga drakii, 255
Phyllophaga forsteri, 255
Phyllophaga fusca, 255
Phyllophaga gracilis, 254
Phytcholoma peritana, 255
Pica pica, 126
Picea, 5, 25, 54, 60, 126, 227
Picea glauca, 56, 81
Picea mariana, 27
Picea rubens, 27
Pieridae, 256
Pin cherry, 46
Pine, 25, 26, 27, 29, 33, 37, 38, 40, 41, 44, 48, 59, 73, 80, 81, 82, 174
Pine barrens, 227
Pine barrens zale, 83
Pine grassland barrens, 3
Pine-oak barrens, 227

291

Index

Pineo Pond, 25, 29, 30, 123
  agriculture, 41
  chronology, 32, 33
  fires, 43, 48
  pollen, 37, 42, 44
  vegetation history, 41
Pineo Ridge, 3, 25-52, 54
  fires, 43, 45
  history, 25
  vegetation, 45
  See also Maine; Mud Pond
Pine/shrub barren, 25
Pink-edged sulfur, 256
Pinus, 3, 25, 73, 174
Pinus banksiana, 26, 81
Pinus resinosa, 25, 80
Pinus rigida, 33, 81, 218
Pinus strobus, 25, 59, 80
Pinus sylvestris, 81
Pinus taeda, 81
Pipunculidae, 244, 245
Pitch pine, 33, 81, 82, 218, 225
Pitch pine/scrub oak barrens, 221
Pityopsis falcata, 6, 18, 20, 129
Plainfield, MA, 199
Planthoppers, 243, 244
Plants, 6, 7
  grassland, 15-23
  rare, 18
  vascular, 16, 16, 21
Pleuroprucha insularia, 256
Plowing, 75, 76
Plymouth, MA, 193, 198, 199
Poaceae, 60
Poanes excaecatus, 257
Poison ivy, 88, 92, 105, 111
Polia purpurissata, 257
Polistes, 258
Pollen, 25, 30, 31, 37-44, 48, 53, 60
Pollen analysis, 25, 27, 29, 30, 32, 48, 53, 56, 59, 61
Polyergus lucidus, 258
Polygala nuttallii, 83
Polygala polygama, 17
Polygonum cuspidatum, 104, 105
Polygonum persicaria, 105
Polyphylla variolosa, 255

Pooecetes gramineus, 6, 120, 141, 175
  in Massachusetts, 187-199
  in Vermont, 202
Poplar, 40
Population trends, 164, 165, 224
  birds, 120, 138, 160-164, 190, 193, 196, 202, 203, 212
  butterflies, 9, 220
  Grasshopper Sparrows, 153-171, 213
  grassland birds, 120, 154, 171-187, 201-210
  insects, 224
  moths, 220
  regal fritillary, 261-275
  sparrows, 194
Population viability analysis, 8, 9, 153-170
Populus, 40
Populus tremuloides, 56
Post oak, 83, 225
Potentilla recta, 105
Poverty grass, 17, 92, 139, 239, 251
Prairie peninsula, 4, 18, 19, 218, 219
Prairies, 17, 19-21
Preservation
  of grasslands, 184
Prickly lettuce, 105
Prochoerodes transversata, 256
Proctacanthus philadelphicus, 257
Protolampra brunneicollis, 257
Provincetown, MA, 191, 193, 198
Prunus pensylvanica, 46
Prunus serotina, 105, 109, 117, 118
Psaphida thaxteriana, 234
Psectraglaea carnosa, 235
Pseudopomala brachyptera, 253
Psilocephala haemorrhoidalis, 257
Pteridium acquilinum, 25
Purple arches, 257
Purple chokeberry, 88, 92
Purple cudweed, 18
Purple love grass, 105
Purple moor grass, 71
Purple needlegrass, 83
Pyralidae, 255
Quabbin Reservation, MA, 125
Quack grass, 105
Quaking aspen, 56, 65

Index

Queen Anne's lace, 105
Quercus, 5, 20, 34, 124
Quercus alba, 82
Quercus coccinea, 82
Quercus ilicifolia, 17, 71, 81, 218
Quercus prinoides, 17
Quercus rubra, 25
Quercus stellata, 83, 225
Quercus velutina, 82
Racemed milkwort, 17
Radiocarbon dating, 57
Ragweed, 40, 60, 105
Ram Pasture, MA, 85-97
Raptors, 137, 148, 149
Rare species, 21
    Massachusetts, 83, 86
    See also Endangered species; Threatened species
Recolonization, 237-250
Records of Vermont Birds, 202, 204
Red admiral, 256
Red-bellied Woodpecker, 126
Red cedar, 105
Red clover, 3, 174
Red fescue, 104, 105, 106, 107, 109, 117, 118
Red-headed Woodpecker, 126
Red maple, 27
Red milkweed beetle, 255
Red pine, 25, 27, 29, 34, 37, 39, 41, 48, 80, 81
Red-spotted purple, 256
Red spruce, 27, 29, 30
Reed, 105
Reforestation, 64
Regal fritillary, 2, 9, 86, 219, 220, 238, 261-275
    breeding, 262, 264
    disease, 271
    habitat, 262, 263, 271
    management planning, 261, 264, 271
    Nantucket Island, 263
    nutritional needs, 272
    population trends, 262, 270
    See also Butterflies
Renia nemoralis, 234
Reproduction
    birds, 149

Resprouting, 73, 74
Restoration
    of grassland bird habitats, 211-215
    of grasslands, 227
    of heathlands, 69-78
    of sandplain grasslands, 79-84, 96
Rhode Island
    Block Island, 15, 220, 262, 263, 264
    Conanicut Island, 264
    regal fritillary, 261, 263
Rhodoecia aurantiago, 235
Rhopalidae, 244, 245
Rhus copallina, 105
Rhus radicans, 105
Rice grass, 17
Risk analysis. *See* Population viability analysis
Rockrose, 18
    See also Bushy rockrose
Rosa multiflora, 105
Rosa virginiana, 88, 92
Rose chafer, 255
Rosy maple moth, 256
Rotational grazing, 75
Rough-fruited cinquefoil, 105
Rough-stemmed goldenrod, 17, 105, 106, 109
Round-headed bush-clover, 17
Roundup, 74
    See also Herbicides; Insecticides
Rubus, 111, 113
Rubus allegheniensis, 104, 105, 117, 118
Rubus flagellaris, 99, 105, 117, 118
Rubus hispidus, 85, 88, 92
Rubus lacinatus, 105
Rumex, 41
Rumex acetosella, 92, 94, 105
Rush, 17
Sable Island, Nova Scotia, 6
Saffron-bordered meadowfly, 252
Salix, 56
Salix humilis, 125
Sambucus canadensis, 105
Sandplain agalinis, 6, 18, 20, 83, 86, 129
Sandplain blue-eyed grass, 18, 83, 86, 91
Sandplain flax, 18, 20, 83, 86, 91
Sandplain grasslands, 3, 79-84, 86, 124

293

Index

definition, 218
Kennebunk Plains, 251
Maine, 238
maintenance, 95
management planning, 81
regal fritillary, 261
restoration, 96
Sandplains
management planning, 217–236
Sandwich, MA, 192, 193, 198, 199
Sanford, Maine, 155–160, 163, 165
Saponaria officinalis, 105
Saturniidae, 234, 256
Satyrium titus, 256
Savannah Sparrow, 6, 141, 143, 145, 149, 171, 179, 180, 191, 208
in Massachusetts, 187–199
in New York, 213
in Vermont, 203, 204, 207
Saxtons River, VT, 205
Scarabaeidae, 254, 255
Scarlet oak, 82
Scentless plant bugs, 244
Schinia bifascia, 235
Schinia lynx, 257
Schinia spinosae, 236
Schizachyrium scoparium, 5, 17, 85, 88, 92, 93, 94, 99, 105, 117, 118, 124, 139, 239, 251
See also Little bluestem
Scleria pauciflora var. *caroliniana*, 83, 93
Scoparia, 255
Scotch pine, 81
Scrub chestnut oak, 17
Scrub oak, 17, 20, 71, 73, 81, 218, 225
Scrub oak plains, 128
Scudderia curvicauda, 253
Scudderia pistillata, 253
Seaside goldenrod, 105, 109
Sedge, 17, 88, 89, 92, 105, 109, 139, 176
Sedge sprite, 252
Sedge Wren, 203, 204, 206, 207
Seed bugs, 243, 244
Semiothisa eremiata, 235
Seral communities, 226
Sesiidae, 255
Shadbush, 6

Sharp-tailed Grouse, 126
Sheep fescue, 85, 88, 92, 94, 105, 109, 251
Sheep laurel, 27, 46
Sheep sorrel, 92, 94
Sheffield, MA, 199
Shoreline, 236
Short-eared Owl, 83, 86, 137, 148, 149, 204, 206, 207, 211
Short-horned grasshopper, 242, 244, 253
Showy goldenrod, 105, 111
Shrubs, 87, 139, 239
as breeding areas, 174
early successional, 6
heathland, 85
in Ram Pasture, 89, 90
Sickle-leaved golden aster, 6, 18, 20, 129
Silphidae, 254
Silver birch, 73
Sisyrinchium arenicola, 18, 83, 86
Slash-and-burn, 122
Slender clearwing, 227
Slender-leaved goldenrod, 105
Small flat-topped aster, 17, 88, 92
Smerithus cerisyi, 257
Snowberry clearwing, 257
Soft rush, 105, 109
Solanum dulcamara, 105
Solidago, 105, 111, 171
Solidago altissima, 17
Solidago canadensis, 105
Solidago gigantea, 17
Solidago graminifolia, 105
Solidago juncea, 17, 105
Solidago nemoralis, 17
Solidago puberula, 17
Solidago rigida, 20
Solidago rugosa, 17, 105, 109, 195
Solidago sempervirens, 105, 109
Solidago speciosa, 105, 111
Songbirds, 211
Song Sparrow, 137, 141, 142, 143, 146, 147
Sorghastrum nutans, 17
Spangled fritillary, 220
Sparganothis fruitworm moth, 255

Index

Sparganothis sulfureana, 255
Sparrow, 120, 131
   in Maine, 137, 140-143, 145, 146-149, 153-170
   in Massachusetts, 86, 187-199
   in New York, 100, 171-186, 211
   in Vermont, 202, 203, 204, 206, 207, 208
   See also Specific sparrow i.e. Grasshopper Sparrow
Spartina pectinata, 17
Spartiniphaga includens, 236
Spartiniphaga inops, 236
Special concern. *See* Threatened species; Endangered species
Specialization, 223
Species decline
   birds in Vermont, 202
   Grasshopper Sparrows, 213
   grassland birds, 188
   sparrows, 171
Species diversity, 94, 221, 222, 223, 251
Species richness, 147
Speyeria aphrodite, 220, 256
Speyeria atlantis, 256
Speyeria cybele, 220, 256
Speyeria idalia, 2, 86, 219, 238, 261-275
   See also Regal fritillary
Speyeria mormonia, 269
Speyeria zerene, 272
Speyeria zerene hippolyta, 271
Sphecidae, 258
Sphex ichneumoneus, 258
Sphingidae, 236, 257
Sphinx drupiferarum, 227, 236
Sphinx poecilius, 257
Spiders, 240, 244, 246
Spiranthes vernalis, 83
Spittlebugs, 244
Spiza americana, 127
Spizella pusilla, 137, 141
Sporobolus compositus, 20
Spraying, 8, 74, 237-250
   See also Herbicides; Insecticides
Spring azure, 256
Springfield, MA, 15

Spruce, 5, 25, 27, 29, 30, 33, 37, 39, 40, 41, 43, 44, 48, 54, 56, 58, 60, 81, 126, 227
Staphylinidae, 254
Staphylinus vulpinus, 254
State listed. *See* Endangered species; Threatened species
Stiff aster, 17
Stiff goldenrod, 20
Stiletto flies, 244
Stink bugs, 244
Stow, MA, 199
Stratigraphy, 36-41, 53
Stream cruiser, 252
Strix occidentalis caurina, 154
Sturnella magna, 127, 141, 145, 175
   in Massachusetts, 187-199
   in New York, 211
   in Vermont, 202
   See also Eastern Meadowlark
Sturnella neglecta, 127
Sunderland, MA, 198, 199
Swamp dewberry, 85, 88, 92
Sweet everlasting, 105, 111
Sweet fern, 25, 27, 31, 41, 44, 46, 48
Sweetfern tubemaker, 255
Sweet gale, 31
Sympetrum costiferum, 252
Syrphidae, 244
Syrphid flies, 244
Tall goldenrod, 17
Tanwave, common, 256
Taraxacum officinale, 105
Temperature, 224, 225
Tenspot, 252
Tephrosia virginiana, 124
Tetraopes tetrophthalmus, 255
Tetrigidae, 253
Tettigoniidae, 244, 253
Thatch, 64
   See also Litter cover
Therevidae, 244, 257
Thinker, the, 257
Thistle, 105, 272
Threatened species, 194, 223
   birds, 15, 119, 172, 189, 203, 211

295

Index

butterflies, 224
  in Massachusetts, 187, 190, 219
  See also Endangered species
Threats, 226
Thrips, 225
Thuja, 33
Thysanoptera, 225
Tickle grass, 105
Tiger moth, 221, 227
Tiger swallowtail, 256
Timothy, 3, 6
Tingidae, 244, 245, 253
Tortricidae, 244, 245, 255
Toxicodendron radicans, 88, 92, 105, 111
Toxostoma rufum, 137, 141
Trifolium, 174
Trifolium pratense, 3, 174
Truro, MA, 193, 198
Tsuga canadensis, 25, 59, 123
Tubuliflorae, 60
Tumbling flower beetles, 244
Tunk Mountain, 29, 30
Turf, 70, 75, 76
Turners Falls, MA, 194, 198, 199
Tympanuchus cupido cupido, 2, 79, 127, 189
  See also Heath Hen
Tympanuchus cupido pinnatus, 128
Tympanuchus phasianellus, 126
Tyto alba, 211
Ulmus, 34
Upland Sandpiper, 7, 8, 120, 126, 127, 129, 131
  in Maine, 137, 141, 142, 143, 144, 147, 148, 149
  in Massachusetts, 86, 187–199
  in New York, 211, 213
  in Vermont, 203, 204, 205, 207
Upland willow, 125
Vaccinium angustifolium, 8, 25, 54, 88, 90, 92, 95, 123, 139, 237, 251
  See also Lowbush blueberry
Vaccinium corymbosum, 27
Vagabond crambus, 255
Vanessa atalanta rubria, 256

Vanessa cardui, 256
Vascular plants, 16, 17, 21
Vegetation, 82, 177
  as breeding areas, 181
  ericaceous, 5, 29, 45, 46
  and fire, 26, 46
  at Floyd Bennett Field, 99
  heathlands, 76
  history, 31, 41
  invasive, 73
  Pineo Ridge, 25–52
  prairie, 18
  and sparrows, 171–186
Velvet ants, 257
Verbascum thapsus, 105
Vermont
  Champlain Lowlands, 201, 202, 205, 206, 208
  Connecticut River Valley, 206, 208
  endangered species, 205
  grassland birds, 201–210
  Lake Memphramagog, 206
  nesting areas, 204
  Northeast Highlands, 208
  Panton, 205
  Saxtons River, 205
  threatened species, 203
Vermont Institute of Natural Science, 202
Veronia, 272
Vesper Sparrow, 6, 120, 129
  in Maine, 137, 141, 143, 145, 148, 149
  in Massachusetts, 187–199
  in New York, 175
  in Vermont, 202, 203, 204, 207
Vespidae, 258
Viceroy, 256
Viola, 224, 261
Viola fimbriatula, 17, 124, 262
Viola lanceolata, 262
Viola lineariloba, 124
Viola papilionacea, 262
Viola pedata, 17, 124, 262
Violet, 17, 124, 224, 261, 262, 264, 269, 271
Virginia rose, 88, 92
Washington County, ME, 3, 25–52, 54
Wasps, 245, 248
Weather, 81, 219, 224
Wellfleet, MA, 193, 198, 199
Wells, Maine, 155, 156, 157, 158, 159, 160, 163, 165

Index

West Bridgewater, MA, 199
Western Meadowlark, 127
Westfield, MA, 198, 199
Westover Air Force Base, MA, 15, 164, 191, 192, 193, 194, 195, 196
See also Airports
Wetlands, 236
White admiral, 256
White birch, 33
White-dotted prominent, 257
Whiteface, 252
White oak, 82
White pine, 25, 27, 29, 33, 37, 38, 40, 41, 59, 80
White spruce, 56, 81
Whitetail, 252
Whorled loosestrife, 139, 239
Wild cherry sphinx, 227
Wildfires. *See* Fire
Wild lettuce, 105, 111
Wild lupine, 18, 20
Willow, 56

Woodbine, 105
Wood nymph, common, 256
Woolly gray geometrid, 221
Worcester, MA, 189, 194, 199
Worthington, MA, 199
Xerothermic period, 20
See also Hypsithermal period
Xylotype capax, 234
Yarrow, 105, 111
Yellow birch, 33
Yellowthroat, Common, 137, 141, 142, 143, 146, 147
Yellow wood sorrel, 105, 111
York County, ME, 9, 239
See also Kennebunk, ME
Zale, 83, 234
Zale curema, 234
Zale obliqua, 234
Zanclognatha theralis, 234
Zerene fritillary, 272
Zizia aptera, 20